On the Laser's Edge
The Conspiracy: Code Word Tikal

by Sharon Thatcher
With Thomas E. Burg, FBI Special Agent, Retired

Copyright 2006
ISBN: 978-0-9795752-0-4

Copyright 2006 Sharon S. Thatcher

All rights reserved. No part of this book shall be reproduced or transmitted in any form or by any means, electronic or mechanical, including photocopying, recording or by an information or retrieval system without written permission except in the case of brief quotations embodied in articles and reviews. For information contact:

Merrill Publishing Associates
P.O. Box 51
Merrill, WI 54452
email: merrillpublishingassociates@verizon.net
www.onthelasersedge.com

The views expressed in this book do not necessarily represent the views of the FBI.

Cover design by Sential Design, LLC
Pagination by Merrill Publishing Associates
Printed in United States of America by DigiCopy, 1800 W. Stewart Ave., Wausau, WI, 54401.

DigiCOPY

Library of Congress Cataloging-in-Publication Data:
Library of Congress Control Number: 2007902992
ISBN: 978-0-9795752-0-4

Second Edition, May 2007

In memory of Mike.

Special thanks to our editor, Ralph Wehlitz,
and to the Muckerheide family (the Schreiners and Barwicks)
for their assistance, patience and friendship.

CHAPTER ONE

When the opportunity to work toward devising a laser afforded itself, the lights suddenly went on and the bells started ringing. The level of my excitement is hard to describe.

— Theodore Maiman, Inventor of the first working laser

Mike Muckerheide could pinpoint the day he fell in love with lasers: July 7, 1960. On the evening news, a bright, thin beam was ignited into life with silvered mirrors and ruby crystal, burst through quartz tube and impaled the tiny slab of a razor blade. Eight-hundred-forty miles away Mike, a 30-year-old lab technician, watched the spectacle flicker before him on a black and white television screen, and awakened to a new life. It was as if that beam had sailed out of the network news studio in Manhattan, crossed over three states, two Great Lakes, the boot of Ontario, and half of Wisconsin just to land inside his brain. Like the world, he had just been introduced to the laser, and neither of them would ever be the same again.

From that moment on, Mike developed one singular passion that would drive the rest of his life: a passion for building lasers. Who would have guessed that a poor boy from the middle of Dairyland would eventually have experiments on NASA space shuttles, own several U.S. patents, work with the U.S. Department of Defense on top-secret military projects, and come to be respected by some of the greatest laser minds in the world...all without a formal college education?

Indeed, Mike Muckerheide was an extraordinary man, driven by the light of lasers; driven, at times, near madness. Yet, it wasn't the lasers that nearly drove him crazy: it was the handful of people who knew, like Mike, what wonderful — and horrible — things this new science promised, and who wanted the knowledge that was locked inside his head. It wasn't just the government, NASA, and great, independent researchers who came calling; there were also the Soviets and the homegrown terrorist wannabes.

September 2004: a Delta Airlines plane was heading for Salt Lake

On the Laser's Edge...6

City International Airport on an evening flight from Dallas, Texas. Five miles from the city, the Boeing 737 started its descent into a routine circular formation in preparation for landing. Suddenly, a flash of intense, bright light blasted through the cockpit windshield and pierced the eyes of the crew. The pain was excruciating. The pilot could hardly keep his hands on the controls. Where had the light come from and would he be able to land the jet safely?

December 2004: news reports began to circulate about a memo issued by the FBI and Homeland Security Department warning that "...there is evidence that terrorists have explored using lasers as weapons." In particular, it expressed concern that lasers might be used to blind pilots during an approach to an airport. Indeed, by the end of the following year — 2005 — there had been nearly 300 verified laser incidents involving pilots in the U.S.[1]

After listening to the news reports, retired FBI agent Tom Burg and I wondered: "If Mike were here, what would he say?" Of course, we already knew the answer, because we had heard it before in the hours of interviews with him. He would have said, "What most people wouldn't believe in the 1970s is all coming true." With a deep-in-the-gut anger he would have added, "I tried to tell them, and they wouldn't believe me."

Mike would wonder what had taken the modern terrorists so long to figure it all out, and what was the government doing to avoid a catastrophe? It had been nearly 30 years, and only now was the world awakening to the horrible possibilities of it: Terrorists using laser beams to blind pilots and air traffic controllers; handheld laser weapons to blind soldiers; and, for the distant future, people, planes and buildings that would simply disappear, as if zapped into a huge Bermuda Triangle. These weren't new ideas. If anyone had wanted a primer, they could have found it in the hidden details of an old FBI case.

Tom knew all the details of the case; at least the exciting parts: about the wire tap at the state capital, the threats on Mike's life, and the list of CIA names confiscated in Chicago from a Wisconsin assemblyman's dinner jacket. Together, we were gathering up all the pieces of the fascinating case, and the equally fascinating character of the self-made laser scientist, to capture it in a book. The FBI/Homeland Security memo had just added one more "I told you

---

[1] "Stiff penalties proposed for laser pointing." *USA Today.* Friday, December 9, 2005, page 10A.

so" to Mike's oddly twisted story.

Tom and I had started on our journey nearly two years earlier.

"I hope he's well today," I said to Tom as he negotiated his car onto Highway 10 out of Stevens Point, heading east. We'd been on the road for nearly an hour and we still had a long drive ahead of us to get to Port Washington.

"Well, you talked to him yesterday. He sounded all right, didn't he?"

"Yes, but I think he's been having a lot of bad days because of the cancer treatments. When he called to cancel our meeting last week I got a little worried. I know this is something he really wants to do, so he must have been feeling pretty bad."

Tom, the eternal optimist, shrugged it off. "I'm sure he'll be fine."

Wisconsin is no land for sissies. Its winter's bite can sting with bitter cold and crippling snows. We picked a good day for traveling. The early January sky was an icy blue, the air snappy, and the roads dry. There was little traffic and we glided effortlessly through small towns and farm country, the congested areas ahead of us in the Fox Valley and the greater Milwaukee area.

To while away the time we talked about Mike, the laser conspiracy case, and his remarkable career. I recalled my first meeting with the scientist. It was 1991, some thirty years into his life with lasers. He was working as Research Director of the laser laboratory at St. Mary's Hospital in Milwaukee. He was a distinguished, medium built gentleman, with a square face and a balding head. The shiny patch of fairway on his crown was surrounded by a half circle of closely cropped hair that was fading from brown to gray. He had bushy, inverted Nike logos for eyebrows, and a mouth that disappeared into a long, thin line. He was very pleasant, quick-witted, and unwaveringly confident.

I remember that we walked down a long, scrubbed-clean hospital corridor into an isolated area, and from there into a large room where a simple contraption of metal and glass occupied the center of my attention. The shape reminded me of a boxy, darkroom enlarger positioned sideways. It sat on a table maybe four feet long, at eye level. Mike told me it was a laser and with apparent eagerness he showed me how it worked. He dimmed the lights, manipulated a switch, and voila! In front of my eyes danced a ghostly image; a hologram of someone — I seem to remember it was Mike — suspended in mid air. It was magical.

On the Laser's Edge...8

For various reasons the article I hoped to write back then about the FBI case didn't work out. Neither did a book project we discussed. Not because we didn't try. Mike mailed me voluminous envelopes full of documents over the course of a year, and occasionally called to see how the book was coming along. "I really need you to sit down and let me interview you," I kept insisting. He had quickly, very colorfully described the case to me on my visit to St. Mary's. Sounding truly frustrated, he always begged off. "We will...it's just, it's just incredible right now. I'm very busy, very busy with something right now." After a year of delay, I decided I was wasting a lot of time getting nowhere, and stopped pursuing the story. I would learn much later that our project had become derailed when Mike took on additional work in secret weapons research for Desert Storm. I wondered if something else would derail our efforts this time.

Mike lived in the beautiful, historic town of Port Washington, on a steep bluff 200 feet back from a dizzying plummet to the waters of Lake Michigan. As if a reflection of his own life, days at his Wisconsin home were blessed with breathtaking beauty and cursed with terrifying storms.

I was apprehensive when we pulled up to his '80s Colonial house. It had been several years since I'd seen him last. He was then a strong and robust individual. Now he was seriously ill and fighting for his life with every ounce of his dwindling energy, trying to accomplish a few more of his endless goals. Would he be the skeletal remains of the man I remembered?

Tom parked in the driveway next to a large motorhome: an uncommon yard ornament, and probably not one appreciated in this upscale community. We had been told that the garage door would be open when we arrived. We should enter the house through there. Mike saw our arrival and called to us: "Come on in."

We followed his voice and found him sitting in an easy chair in the living room angled into a corner near patio doors that looked out over a wide expanse of lawn that seemed to float off and into the distant water.

Mike motioned for us to take a seat. I positioned myself directly facing him in an upholstered chair near a bookcase filled with scientific manuals and scrapbooks of professional memorabilia sharing company with a bottle of holy water and a bust of the Virgin Mary with a tear in her eye. Tom stationed himself on a nearby sofa beneath framed displays of NASA commendations and space exploration souvenirs.

There was humor in Mike's voice despite all he had been through. In just the past year-and-a-half, he had endured the death of his beloved wife and his younger sister while fighting his own long and difficult battle. He was getting aggressive treatment, however, and was still living alone. He talked optimistically about the laser research he was developing at a small, private lab nearby.

Mike had aged gracefully, I noticed, despite his condition and the emotional challenges of the past year. Though he had to hobble to his destinations with the help of a walker, he did not appear as frail as I had anticipated. In fact, at 72, he still cut a solid, imposing figure with a personality to match. He possessed a no-nonsense drive to keep plunging ahead. He was refusing strong pain medication to keep himself alert enough to finish his various projects. As well, he had dismissed the daily oversight of hospice nurses because "all they wanted to talk about was what kind of casket I should buy," he said, adding, "I know where I'm going. I don't need to talk about it all the time."

What he did want to talk about were the events of his life, his career, and the FBI case.

Tom and I had barely settled in when Mike began to tell his story. For the next four hours we had a hard time keeping up with the rambling dialogue.

"In 1969 I went to the Pentagon to meet with them. I figured out how to use anthrax and bubonic plague in the mail... I gave it all to [Secretary of Defense] Melvin Laird. Then I went to meet with this guy named Emerson... And Emerson gave me this letter to give to a guy named Wilson at the University of Wisconsin, but I never did because when I got back home my lab had been broken into and the cops who looked at it said, 'This is not normal the way this was done'."

Before we knew it, Mike was veering in a new direction. "I can't remember why this happened, but I went to meet with one of these Posse guys in a thicket, and I was full of ticks when I got home. So I took a shower, a cold shower, to try to get them off."

And then: "I went to an attorney friend of mine with the intentions of getting a patent for this device to protect policemen and firemen. Today it's called a dazzler, and for some reason this attorney wrote to J. Edgar Hoover about it and he got a letter back from J. Edgar Hoover saying he was interested but there wasn't any money to support this technology. Then about two years later a couple of them with money went to the feds and they got funding. But I wasn't big enough."

To our relief, about midway through, Mike's cockatoo Peppy began to sing from his cage for the first time since our arrival, and temporarily sidelined the oration.

"My wife Pat used to come down in the evening in a robe and he'd go, 'Pretty Girl,' and whistle. He has a 38 word vocabulary and has a song he sings, *Mary Had a Little Lamb*, and he gets it all mixed up."

It's just a breather, though, and he's off and running again.

"I found the first letter I ever wrote to NASA the other day...."

Mike was like an overturned vessel, with memories flowing out like an uncontrolled oil spill. My handwritten notes were an indecipherable mess of scribbled names and technical terms. I was relieved to be using a tape recorder, but constantly worried that it wasn't doing its job, especially since Mike's words frequently slid off into a whispered mumble. Was the microphone close enough? Did the machine quit when I didn't notice? Did I remember to push the record button when I changed tapes?

After two days of interviews, Tom and I began to realize how extensive the project ahead of us would likely be. Not only was the case itself complex, but so was Mike. He would be labeled a genius by many of his friends and associates, not because of what he knew, as much as for how he knew it: lacking any degrees beyond a high school diploma, he had found a respected place in the laser field through trial and error, not overlooking heavy doses of self drive and hard work. His devotion to Catholicism, family, country and career had been molded by a lifetime of experiences still to be understood. "I feel like we're trying to put our arms around a gorilla," a rather dazed-looking Tom said at the end of our visit. And in time, we knew the gorilla would grow.

Knowing that Mike suffered ill side affects from his treatments, and not knowing his treatment schedule, I suggested that he call me whenever he felt like talking. The calls began about three weeks later.

The first one came as a pleasant surprise: I hadn't known if he would follow-up or not. Even more pleasant was listening to him. Unlike the man Tom and I had visited in Port Washington, whose words were pinched from physical discomfort, Mike's phone voice claimed a calm, airy quality. Since he always called me on days when he was feeling good, I came to know him as someone with great humor and genuine concern. Across the miles of fiber-optic cable, years were peeled away to reveal his boyish enthusiasm. It was the voice of the young Mike, the cancer-free Mike, who called.

We talked every two or three weeks. I always tried to be prepared to pull a thread of thought from our original interview, knowing that all the facts were there and just needed coaxing into clarification. I wasn't always successful since Mike had his own agenda for calling. Sometimes he wanted to discuss something relevant to our discussions that he had just seen on one of the news channels, or to add some little fragment of personal history he'd suddenly remembered. In some ways, it was like digging up soil that hadn't been disturbed in years: sometimes a kernel of detail would mysteriously pop loose from its dormancy and grow something interesting.

CHAPTER TWO

*It is still a matter of wonder how the Martians are able
to slay men so swiftly and so silently.*
Many think that in some way they are able to generate an intense
heat in a chamber of practically absolute non-conductivity.
This intense heat they project in a parallel beam against
any object they choose by means of a polished parabolic mirror
of unknown composition, much as the parabolic
mirror of a lighthouse projects a beam of light.
— H.G. Wells, War of the Worlds, 1898

You can't blame the first FBI agent who ever interviewed Mike for being a little skeptical.

"Now Mike, let's go over this again," Agent Jerry Southworth was saying. He was seated at his desk at his office in the federal building in Wausau, Wisconsin, scribbling notes on a yellow legal pad. Mike was sitting across from him. "These guys want you to build this thing inside an airplane."

"Yes, in the nose," Mike responded.

"And this is like a gun, but not a gun."

"Well, not a conventional gun. It's a laser."

"Right. And it shoots...what?"

"Light. A beam of light."

"Okay. It shoots light."

"High powered light. You know, like Buck Rogers. Flash Gordon. It's not fiction anymore, Jerry. It's real."

Mike was perceptive and he knew the agent was having a hard time believing him. Almost everybody did, especially when he started in with the Buck Rogers and Flash Gordon metaphors. He liked the agent, though, and he was willing to be patient. Besides, what other choice did he have? What else could he do when a couple of nutty extremists wanted him to build a laser weapon in the nose of an airplane so they could go out and blind people?

Code Word Tikal...13

"They want to target O'Hare Airport in Chicago. They've also mentioned some Jewish synagogues, and Castro's Havana Harbor," Mike plowed ahead.

Southworth shook his head, but put on his best game face. By now he was thinking: *Either this guy is one of the craziest men I've ever met, or one of the smartest.*

If the conversation had taken place three decades later in history, Southworth would have called Homeland Security and a bevy of black suits would have arrived to take Mike away, maybe put him into a witness protection program, and arrested a couple of bad guys. This was the mid-1970s, though, and Southworth was more inclined to call the men in white suits: the ones with the straightjacket wardrobe and the keys to the mental health center. Something stopped him, though: the inclination to believe Mike.

The agent had good reason for offering allowance. Somewhere between McCarthyism and the John Birch Society, the Ku Klux Klan and Skin Heads, a new crop of ultraconservative extremists had sprouted in the U.S. and was planting itself in Southworth's territory. The group was an obscure, militant offshoot of the Christian Identity Movement called the Posse Comitatus. It was several years before its most notorious member, North Dakota tax protester Gordon Kahl, thrust the group out of obscurity with two bloody shootouts that resulted in the death of two U.S. Marshals, a county sheriff, and Kahl himself. Southworth already knew, however, that when one of its supporters routinely used words like *weapons*, *planes*, and *attack* in the same sentence, it sometimes paid to listen.

It all started innocently for Mike with a knock on the Muckerheide door. It was the early 1970s, and Mike and his wife Pat and their young daughter Susan were living on the southern edge of Wausau in the community of Schofield. It was a quiet town, yet to see the area's spiraling growth that would mark the 1980s and beyond. Despite Mike's evolving success as a laser scientist, he was still struggling for financial success, and the family lived in a comfortable, but modest home.

On the other side of the door that day was local investment counselor John Claussen. He was a tall, smiley man in his late 20s with a long, thin face and deep-set eyes, who dressed like a country boy sent off to the city without style instructions. He had compromised the discomfort of his suit with a pair of oversized rubber galoshes with metal buckles that rattled when he entered, and a tweed, Sherlock Holmes-style hat with ear tabs.

Mike liked to see a young man with ambition, and the visitor came with references, so he invited him in. That was the wrong thing to do.

An acquaintance once described Claussen as "like something stuck on the bottom of your shoe that you can't get off:" an apt description for a persistent salesman. He continued to come back to the Muckerheide home.

Mike insisted that he never invested with Claussen and did not welcome the visits. "I didn't have anything to invest back then," he said. "Besides, I didn't trust him. He came across to me as kind of sleazy."

It begged the question: Why did Mike keep opening the door to a salesman he didn't trust, inviting the guy in and talking to him, when most people would have either slammed the door in his face or pretended they weren't at home?

At first, Mike said it had to do with the conversation: when the sales pitch failed, the conversation turned to politics, and Mike liked talking politics. He had strong opinions, especially about the war in Vietnam, which was then winding down to an ugly conclusion. "The government didn't really give a rip about those men. They were dying, and it was all politicized," he believed. "Men were trying to bomb North Vietnam, and couldn't. The government was restraining those pilots. There was a lot of garbage and I was convinced [the government] was corrupt."

Claussen had his own strong opinions. He was very conservative, very religious and very principled, but in a way even a conservative, religious, and principled guy like Mike found extreme. To an inquisitive-minded scientist, it seemed interesting. More importantly, their paths crossed all too often in the small community; the two men became acquainted by chance, and it's hard to turn an acquaintance away.

Still, Mike wasn't the type inclined to be wasting his time on guys he considered sleazy, all for the sake of a conversation. What really piqued his interest were Claussen's repeated references to the Posse and the politics of the radical right. Sometimes it wasn't clear if he was there to sell investments, or memberships to the political movement. Mike had read and heard enough about the Posse to believe they were misguided and dangerous. More curiously, a name kept surfacing in their conversations about a local chemist who supported the cause and liked to experiment. To a fellow scientist, who had himself experimented with deadly chemicals, it sounded like a dangerous combination.

He likely would have found a way to dislodge the pesky salesman from his life if not for what happened next. When Mike contacted the FBI on a professional matter related to his laser research, he mentioned Claussen's name. Agent Southworth recognized it immediately. The investment counselor was already on his radar screen as someone who liked to brush shoulders with Posse members. He jumped at the chance at getting more information.

"I had the Posse case. The more intelligence you have about that group the better off you are," Southworth would later admit. "When I was aware that Mike knew Claussen, it became a logical thing to do to encourage that association." He hoped Mike would continue meeting him.

The Posse Comitatus had started setting up camp in the eastern part of Marathon County, and the western part of the adjoining county of Shawano, in the early 1970s. Part of a small, loosely organized national organization, it was named after an old law of posse comitatus that allowed for a posse of citizens to help establish law and order. The group's platform was a mongrel mix of religion and politics. On the religious side, they preached that the U.S. would have fewer problems if it had fewer Jews. Jews, too many of whom were manipulative bankers, held too much control, they said, and were conspiring with the United Nations to create a new world order. God's chosen people were of Aryan descent and they had the right to take the federal government back by force.

Potential members who didn't buy into that had a more popular political choice: they advocated no taxes, minimal federal control or licensing, and no authority higher than the local sheriff: it wasn't mainstream, but it had its appeal.

Restricted by how the FBI could investigate such political groups, Southworth could only depend on sources like Mike to come in and volunteer information. It was a legal way of keeping tabs on people of interest which he might not have been able to do otherwise. Mike was willing to keep meeting the quirky investment counselor just to see if anything developed. It finally did, on March 31, 1976.

Alabama Governor George Wallace was making his third attempt

On the Laser's Edge...16

at running for president and was scheduled to be in Wausau for an afternoon rally. Although the controversial governor was running as a Democrat, he was the type of candidate who was popular enough to make the ends of two opposing, political parties meet. Many had held allegiance to him since he first ran for the office in 1968 on the American Independent Party ticket on a platform of renouncing taxes and government interference. Many Posse members had started out as American Independent Party[2] members and were still cautiously supportive of Wallace, believing that he had found it necessary to trade parties to work deals with the devil, and that he would work for their ideals once elected.

Claussen had invited Mike to the Central Wisconsin Airport to wait for Wallace's arrival. "We may be traveling in the Wallace motorcade to the banquet hall," Claussen said, obviously excited about the prospect. Several Posse members were expected to attend the event.

Mike didn't support Wallace or the Posse, and he didn't particularly like Claussen, so he really didn't want to be at the airport to see any of them. He wanted to be in his lab with his lasers, or maybe at home with his wife and daughter. He was there to help Agent Southworth.

Okay, so maybe he was a little curious, too.

"There he is," Claussen said, nodding towards a man in the distance. It wasn't Wallace he was nodding at, however: the candidate's plane had not yet arrived. This was the man Claussen talked about with near reverence in his many conversations with Mike. "You two would really hit it off," he often told him. "The guy's a genius, really intelligent. You'd have a lot to talk about."

Mike had his doubts. Albert Iwen had started down a path to an illustrious career, at Boeing and Brown Engineering, when he suddenly tracked off course into political extremism. He was now a self-employed chemist at a private laboratory in nearby Mosinee and the Feds were curious about what he was cooking up there.

The man approaching was about five-feet-ten-inches tall with tiny, dark eyes rimmed by large, heavy glasses and a prominent nose. Like Mike, he would celebrate his 46th birthday in April. He was dressed, as he was accustomed to, in a conservative suit and tie. Some people would compare him physically to the cartoon character Caspar

---

[2] The American Independent Party was later renamed the American Party.

Milquetoast. One of his attorneys would describe him in court documents as "sort of the absent-minded professor, who tends to ramble sometimes when he talks."

Claussen made the introductions. "Mike, this is the guy I told you about: Al Iwen. Al, this is Mike. He's a scientist too."

The two strangers shook hands. "Pleased to meet you. I understand you're into building lasers," Iwen said.

"Yes. That's right. I work for a research foundation in town."

"I know. I've read about you. I read that paper you did for publication, on lasers and x-rays. Fascinating stuff."

Knowing a little about Iwen ahead of time had given Mike a head start on being suspicious. Now, he was getting paranoid. Though his work was occasionally publicized in the local newspaper, the article, "Laser-generated Plasmas as a Source of X-rays for Medical Applications," compiled by Mike and seven other researchers, had appeared in a technical publication in 1974. To most people it wasn't recreational reading, and the *Journal of Applied Physics* wasn't something a person would find on sale at a local magazine stand. It would have taken some dedicated research to find it; so, why had Iwen bothered? When Iwen continued to mention obscure details about Mike's career, Mike could only suspect Claussen had tipped him off and he was surprised by how much information had been sucked in during their visits.

Nearly an hour passed and still Wallace's plane did not land. Other men at the airport came and went in the conversation, but whenever possible, Iwen pumped Mike for information about lasers. Finally, he motioned him off to one side and handed him a business card. "I'd like to talk to you," he said. "If you have any idea of special force fields in your kind of work, I'd be interested. I know a foreign country that's extremely interested, and there's lots of money."

The card listed the name of Iwen's research laboratory. Mike took the card and said, "Sure. I'd be willing to talk to you."

Mike couldn't leave fast enough. Wallace's plane, 90 minutes late, was finally about to land and the candidate would be whisked away to a reception of about 400 people, but there were other things on the scientist's mind. During their conversation, Mike noticed that the rebel chemist had not been particularly interested in the potential medical applications for lasers. Most of his questions had been directed towards their potential military applications. Mike had been pur-

On the Laser's Edge...18

sued in the past by questionable people who wanted that kind of knowledge. He gave a convincing excuse for leaving and headed directly for a telephone.

Southworth was at his office when the call came in. It was late afternoon.

"Hello, Jerry?" Mike began.

"Yes."

"It's Mike Muckerheide."

"Hi, Mike, what's up?"

"I went to the airport to meet George Wallace this afternoon. I met Iwen instead. I finally met Iwen, and..." He hesitated. "Jerry, I think he may be up to something. I mean something bad, something really bad."

Mike sounded breathless, nearly hyper, and Southworth recognized that something was wrong. "There's a park off Sturgeon Eddy Road," the agent instructed. "Memorial Park. Do you know where I'm talking about?"

"Near the hospital?"

"That's the one. I'll meet you there. Half an hour?"

"I'll be there."

Southworth gathered the litter of papers from his desk, shuffled them into a folder, then locked the folder away in the office safe. He donned a jacket to protect himself from the chill of the early spring air and called his wife to let her know he would be home soon. "I just have one stop to make first," he told her. "It shouldn't take long." He hoped, for once, he was right.

He locked the office door behind him as he left, went down the stairs, and out into the small parking lot.

It was only a five-minute drive to the park and Southworth knew he would arrive early, and he wanted it that way. It was an agent's trick to set and control every scene as much as possible, even if the situation seemed harmless.

The drive took him past the historic brick buildings of downtown Wausau, the new courthouse, and the elegant, Victorian homes along Grand Avenue. At the intersection of Grand Avenue and Sturgeon Eddy Road stood Wausau Hospital. Southworth turned right at the hospital, and proceeded down Sturgeon Eddy Road, negotiating his Bureau car around a left-angled bend before reaching a driveway marked *Memorial Park*. It was one of many small neighborhood parks tucked into the landscape around the budding community.

Southworth turned right at the entrance and quickly slipped out of view as the driveway sloped steeply and turned. At the bottom was a wide area of trees and manicured lawn nestled against the shoreline of the Wisconsin River.

With evening closing in, the park was quiet. Seeing nothing of concern, the agent parked facing the entrance and settled in for the wait.

Minutes later, a car came cruising down the drive and into the parking lot. It was Mike in his white Pinto station wagon.

Mike pulled alongside the agent's sedan, driver to driver. The two men rolled down their car windows and turned their ignition keys off.

"So, Mike, what happened?" the agent started.

Mike was worriedly shaking his head and darting glances off into space. "Jerry, it's this guy Iwen. I don't trust him. He's talking special force fields and selling to a foreign country."

"What do you mean?" the agent coaxed. He didn't understand what Mike was talking about when it came to special force fields.

"You don't sell this stuff to foreign countries. It's serious technology."

"Lasers?"

"Yes. Lasers. Powerful lasers. Dangerous lasers. You're dealing with something that's hotter than the sun. They can heal in an instant and destroy just as quickly. Lasers are going to revolutionize this world, Jerry. That's what Iwen wants to sell away to another country."

As interesting as it all sounded, nothing illegal had happened and Southworth knew that he couldn't do much with the information unless something else did happen. He calmed Mike the best he could. "You know, Mike, most of these guys in extremist movements are all talk. They blow off steam and go home. I doubt if he'll do anything, but we'll keep an eye on him, and I do appreciate the information. Call me any time."

Although the agent truly believed what he had just said, he was honestly a little confused and concerned by the scientific spin of Mike's story. Since arriving several years earlier to reopen the Resident Agency of the FBI at Wausau he had gotten to know members of the Posse Comitatus too well. They were not opposed to making their point with the end of a loaded gun, but the added dimension of laser science just didn't add up to typical Posse tactics.

On the Laser's Edge...20

Southworth had to wonder about his informant. This wasn't the first time Mike had been connected to a strange story. Only a year previously the scientist had walked through the doorway of his office and handed him a letter from a professor in the Department of Theoretical Physics of Kaliningrad University in the Soviet Union. The letter requested a reprint of an article Mike had published in a scientific journal:[3] the same article he said Iwen had mentioned. Mike had thought the Soviet request strange since he felt they obviously had access to the publication. "If all they want is a copy," he rationalized, "Can't they get it themselves?" He thought they might be fishing for something else: perhaps a scientist willing to sell valuable, new technology.

It was years before the iron curtain would fall and the Soviet Union and United States were still Cold War enemies. The request was unusual. Southworth questioned why the Soviets would be interested in a small town scientist, but he could take no chances. With direction from FBI Headquarters in Washington D.C., he helped draft a three sentence response, [4] had Mike sign it, then mailed it off across the Atlantic. The Soviets had not responded.

Southworth remembered yet another odd encounter more recently with the scientist. Joined by several observers, Mike had been experimenting with a laser up in the 3M granite quarry located on the west side of nearby Rib Mountain. The local police had been properly notified in advance about the experiment, but when reports came in from the public about UFO sightings that evening, they began to wonder if the lasers might cause contamination. They decided they needed to be better prepared with some answers and called the FBI for assistance. It was a quiet investigation that uncovered no hazards, but for the first time it raised questions in Southworth's mind: who was this character, and what was he up to? He would go back to his office, write up his report, and file it away, hoping that Mike would never contact him again, and fearful that he wouldn't.

---

[3] "Laser-generated Plasmas as a Source of X-rays for Medical Applications" by P.J. Mallozzi, H.M. Eptsein, R.G. Jung, D.C. Applebaum, B.P. Fairand, W.J. Gallagher, R.L Uecker, M.C. Muckerheide, *Journal of Applied Physics*, April 1974.

[4] Copies of the Soviet's request and Mike's response from the Muckerheide files.

CHAPTER THREE

*Laser warfare will be different than anything
mankind has ever seen.*
— Mike Muckerheide

On January 4, 2005, the FBI made their first arrest in the rash of laser incidents involving pilots. The man arrested admitted that he flashed the laser at two different aircraft on two different nights: the first, a small charter jet, and then a helicopter (carrying, no less, Port Authority detectives who were out looking for him). The accused, David Banach, said that he pointed the light beam at the two aircraft from his back yard near Teterboro Airport in New Jersey using a handheld laser device readily available off the internet and purchased for his job testing fiber-optic cable. Why he was pointing it at aircraft and not fiber-optic cable was not clear (pointing out stars, his lawyer said) but authorities agreed that he was not a terrorist, just a "foolhardy and negligent" cable guy.

That might be the end of the story, if not for the thought that haunts: if an average-Joe can do this with a $100 handheld laser, what do we face from terrorists bent on real destruction?

It wasn't a question that Agent Southworth had the history to ponder back in the 1970s. He met Mike and the imaginative Iwen three decades before lasers were so easily available to star gazers, cable searchers and others of the general public. Very few people yet understood them or their potential, and Mike's attempts at educating the agent weren't working. In a desperate, almost giddy way he would try to explain how eyesight would one day be restored in seconds; how wrinkles would be erased to reveal youth; how internal surgery would be performed through tiny holes in the skin and result in miraculous recoveries. Alas, it fell on skeptical ears. Sure, Southworth conceded, maybe it could work; after all, he had read the occasional newspaper articles, but Mike's contentions that a person could take these nice, friendly lasers and turn them into blinding

On the Laser's Edge...22

machines — maybe even killer machines — seemed a little far fetched.

Mike could sense the doubt. Regardless, he liked the guy. Most people did. The former Michigan police officer was amusing and took a common sense approach to his work that made him a favorite among his peers and superiors. Mike called him "swashbuckling." It came from the agent's harmless rebellion to convention: his preference for dressing down in colored shirts at the office when the standard was white shirts and dark suits, and the growth of a mustache which he thought necessary "to make myself look older:" an early acquisition when the Bureau relaxed its standards.

His signature piece was a wide belt: a holdover from his police days, because it hung better when handcuffs and a holstered gun were attached. He joked that he kept an extra notch in the belt so he could tighten it up when getting ready for the periodic weigh-in's mandated by the Bureau, then let it out when they were over.

Mike had started to meet on a regular basis with both Iwen and Claussen after the George Wallace rally. Every couple of weeks the trio would meet for breakfast or lunch at one of the local restaurants where the conversation would inevitably drift into politics and government corruption. Mike would sit and listen carefully to his new "pals," mentally taking notes, then duck off to a clandestine meeting with Southworth.

Bureau policy recommended that agents meet with their sources someplace other than an FBI office, such as a public park. In this case Southworth's office was less visible in the small community and Mike was often instructed to go there. It was located inconspicuously on the nearly deserted second floor of the three-story, cream-color brick federal building downtown. The building also housed the Internal Revenue Service and the district office of Congressman David Obey. One could easily walk through a back entrance unseen, but if seen, Mike could have invented a good excuse for being there that would have concealed his true destination.

At first, the conversations with Iwen and Claussen were nothing much out of the ordinary. It was a time of growing skepticism in the country and even Mike had his complaints. It had been a tough couple of decades for the U.S. First, there had been serious reservations over the country's handling of military intervention in Korea, followed by the unexpected assassinations of President Kennedy, his politically aspiring brother Bobby, and the admired civil rights leader Martin Luther King. Add to that the many years of race riots and war protests, and deep wounds were left in the American psyche.

Public opinion offered a mixed bag of solutions. Mike believed in the overall integrity of the system and the power of the vote. On the extreme edges were people who believed the government was hopelessly corrupt, and only by taking the power forcibly away could freedom be restored. Claussen and Iwen had chosen the latter.

As the three men became more accustomed to meeting, it was abundantly clear to Mike that Iwen had bought into the Posse platform: that there was a secret government operating at the federal level which was controlled exclusively by Jews. Revolution was necessary to break that power and he often suggested bombing Jewish properties in the U.S. to accomplish the goal. "He wants to change this country. Bring it under different rule," Mike reported. "He has ideas from Hitler's boys: Deutschland garbage. It's a mixture of all kinds." To show his allegiance to the cause, Iwen often folded back the lapel of his suit to show off a pin in the shape of a swastika.

If the racist and anti-Semitic comments Mike routinely heard at these meetings bothered him, it was only eclipsed by Iwen's preoccupation with weapons and his flippant suggestions at destruction. Was it just talk? Mike didn't think so. Southworth wasn't sure. One of the local Posse leaders was already in trouble for assaulting an IRS agent. Iwen was not your average extremist; he was intelligent, focused, and obviously disgruntled about something — just the type the FBI wanted to keep an eye on.

Iwen had hinted at inviting Mike to his laboratory in Mosinee. When he finally did, Mike asked Southworth for his advice: "Should I go?"

"Sure, if you're comfortable with it," the agent replied. Secretly, he was happy to have an opportunity at finding out what was really going on there.

The lab had originally been a legitimate dental supply house, then, in 1973, a likable but controversial pharmacist took ownership. Soon the company, Mosinee Research Corporation, listed the pharmacist, David P., as Director, Albert Iwen as President and General Manager, and a third man, Thomas Stockheimer, as an employee. All had connections, past or present, with the Posse Comitatus and other extremist movements; in particular, Stockheimer who had founded

On the Laser's Edge...24

the Wisconsin movement. Stockheimer was also the one who had assaulted the IRS agent, and he had recently failed to show up for his scheduled paid-vacation at federal prison. He was now listed as a fugitive.

Mike drove himself the twelve miles to Mosinee for the visit, stopping at a long, squatty building on Main Street with a front of plate glass windows. Despite looking like the Red Owl grocery store that it had once been, the inside revealed a total conversion: office space in front and a state-of-the-art laboratory in back, with work areas of stainless steel awash in bright, florescent light. From the beginning, Mike overheard employees worrying about wire taps on the phones and electronic surveillance bugs in the building: strange, but not as strange as Iwen.

As he was led on a tour, being introduced to workers, Mike listened as the chemist focused on lasers. It was all technical, scientist-to-scientist, matter of fact. "He asked me about separating uranium with laser in special centrifuges," Mike said. "He had a real bent towards that level of the technology." In the Dr. Jekyll and Mr. Hyde of lasers, this was the Mr. Hyde.

As his tour guide babbled on, Mike could feel his own world starting to tilt.

"Something to put into the nose of an airplane," Iwen was explaining. "There's a pipeline to build it and to get it out of the country."

Mike was feeling overwhelmed: like he was venturing too far, wading in too deep.

Iwen's voice continued to echo in his head: "We could blind the controllers at O'Hare Airport. Blame it on the Cubans. Start a war.... Turn the country around..."

Then Iwen paused. "While you are doing something in this lab for me," he said cautiously, "I want you to have something on me."

Mike followed him into a back room where he was hit by the strong, pungent smell of ether or alcohol. He quickly adjusted as the odor drifted off through the open windows.

There were a couple of men in the room, working hurriedly around large, magnetic stirring devices. Iwen explained that they were working with a substance called amygdalin: a white crystalline glycoside that occurs in the kernels of fruits such as almonds, apricots, cherries and peaches. In this case, they were making laetrile, a controversial drug being touted as a cancer cure. It had not been approved by the Food and Drug Administration, and was seen as a get-rich quick scheme [5] spreading across the U.S. from Mexico.

Mike knew about the drug and the controversy. Iwen had talked about it before, without admitting any first-hand knowledge. Mike saw Iwen's revelation as a rite of passage: Iwen trusted him and was willing to seal their new partnership with the secret of his lab, and somehow, it was all connected to a more elaborate plan for lasers.

Mike was hoping that the laetrile operation would be enough to get Iwen arrested and out of his life. Back at the FBI office, however, Agent Southworth had a lot of questions.

"Tell me, Mike, what is it that Iwen wants?"

"He wants me to build a laser system that can be put into an airplane and flown over a group of people, blinding everybody. He says he has a pipeline to build it and to get it out of the country."

"Is that possible?" the agent wondered.

"Very possible."

"So, what did you tell him?"

"I told him the truth: I could do it."

"You can?"

"Yes, I can."

Southworth tried not to look so incredulous. "But you won't, right?" he ventured carefully. He wasn't yet convinced that Mike had the scientific ability to develop such a device, but the possibility nagged him.

"Of course not," Mike assured him, not knowing, so early in the game, that he would ever consider anything else.

Although Mike believed that there was a connection between the laetrile and lasers, Southworth routed the information on laetrile to the Food and Drug Administration (FDA). The illegal manufacturing of a drug came under their jurisdiction. That would at least partially

---

[5] Laetrile is considered one of the most controversial drugs in U.S. history. In 1977 it was banned from interstate commerce after the Food and Drug Administration cited an absence of scientific evidence that the substance was safe for human use. The ban was upheld by the U.S. Supreme Court in 1979. The results of a study sponsored by the National Cancer Institute were reported in the *New England Journal of Medicine*, January 1982, concluding that laetrile was not an effective treatment for cancer and was harmful. The internet, however, has revived a market for it in recent years, and today it is also sold as vitamin B-17, madelonitrile and amygolaloside.

On the Laser's Edge...26

clear his desk to focus on Mike. His first thought about the laser scientist, left unspoken, was: "This guy's a space ranger. He's really out there."[6] It was time to delve deeper into the curious character who had become his informant.

He enlisted the help of FBI Headquarters. Documenting everything, Southworth compiled a summary of what he knew about Mike's research work and some of his conversations with Iwen. Almost tongue-in-cheek, he described what kind of laser Iwen supposedly wanted Mike to build.

He mailed his report off to Washington, D.C. half expecting to learn that Mike was just a run-of-the-mill, Buck Rogers wannabe. Instead, they shot back their response: Keep investigating; this guy knows what he's talking about.

[6] Interview with Jerry Southworth

CHAPTER FOUR

Do not underestimate the intense drive and motivation of a maverick scientist.
— *Theodore H. Maiman, The Laser Odyssey*

It is understandable why Southworth was prone to question Mike's background. Long before the real laser was created, its evil, fictional twin, the ray gun, had established a bad reputation at the hands of cartoonists and science fiction writers. Once the laser was introduced, people immediately recognized the similarity.

Theodore Maiman didn't like the comparison. He hoped the science would be used for the good of mankind, not the bad, and blamed the media for jumping to the wrong conclusions. He said it was a *Chicago Tribune* reporter who asked him at a press conference at the laser's debut if it was going to be developed as a weapon.[7] Maiman stammered to answer, finally admitting that it *could* be developed as a weapon, triggering foreboding headlines around the country: "L.A. Man Discovers Science Fiction Death Ray."[8]

Despite its bad beginning, the laser did get a jump-start on winning the public over. The technology of making good lasers was easier than making them bad, and many wonderful and miraculous, medical advancements soon followed. Lasers also moved successfully into the public marketplace through a myriad of household and workplace gadgets that made life easier, from surveying equipment, to price scanners, to printers, and video equipment. For the laser, it was a great public relations turnaround.

The word laser is an acronym for *light amplification by stimulated emission of radiation*. The door to its discovery was opened in 1911 by Danish physicist Niels Bohr with his quantum theory of how

---

[7] *The Laser Odyssey* by Theodore H. Maiman, Laser Press, 2000
[8] *Los Angeles Times*, July 8, 1960.

On the Laser's Edge...28

atoms emit energy. Albert Einstein followed with his theory of how atoms and molecules absorb and emit radiation. A few years and a few scientists later and all the principles were in place to begin transforming the science into hardware.

There is, and will likely continue to be, controversy over who should have been awarded the patent rights to the first laser. After a very lengthy patent war that didn't end until 1987, the honor went to Gordon Gould. He was a student at Columbia University when he developed a written concept on lasers. In the legal fray also was the team of Charles Townes and Arthur Schawlow. Townes and Schawlow had advanced their theory on lasers based on Townes' earlier invention of the maser, which utilized a similar principle using the less powerful microwave frequency.

With patents, timing is everything. A crafty lawyer with persistence made all the difference in the laser patent war.

In some kind of poetic justice, however, Nature settled the question once and for all; it was discovered that lasing (what lasers do) occurred naturally in the Martian atmosphere.

After that, it was only man-made lasers that could be argued about, and as a don't-just-talk-about-it, get-it-done kind of guy, Mike's allegiance was always to Maiman. He was given full credit for building the first working laser.

The Maiman laser that much of the world saw in the print media at that time, however, was a larger version that a photographer insisted made for a more dramatic shot. The first one could fit into the palm of the inventor's hand. Mike saw it once on a visit to Maiman's home. "He nearly dropped it," Mike said. "He carries it around with him. He was taking it out of his pocket and one part almost dropped on the marble floor and he said, 'Oops, I shouldn't do that.'"

The extraordinary light of laser is often compared to ordinary light. In ordinary light, the atoms of the beam take a leisurely, somewhat random path to illuminating things, going this way and that. The atoms of a laser, however, are stimulated to a much higher degree; instead of spreading out, they line up purposefully in a straight, unwavering line, heading to one destination with incredible speed and precise aim.

When Mike occasionally visited a local school to explain to young students what a laser was he would first draw several squiggly rays shining out of a ball of sun and out of a light bulb. He would explain that light waves go out of the sun and out of the light bulb in all directions and in different wavelengths, bouncing off things and reflecting back. When the light reflected off an object into the eye, we were able to see the object.

Then he would draw two mirrors, parallel to each other with a hole in the center of one of them. He would now draw straight lines — beams of light — back and forth between the two mirrors, each line perfectly straight and perfectly proportioned, clones to one another. And finally, he would draw one beam escaping through the hole.

"So," he would conclude, "a laser is a device with a mirror on each end that reflects light back and forth until eventually it exits one end in a focused, consistent, wavelength creating a laser beam."

He would explain that there were several types of lasers — solid, fluid, gas, and semiconductor [free-electron and x-ray lasers are newer additions] — and within each type a multitude of specific mediums used to stimulate the beam, such as krypton, carbon dioxide, ruby crystal, and so on. Some lasers were so small they could only be seen with a microscope while others were hundreds of feet long. They all projected beams traveling at the speed of light. They were also extremely hot, but in a very concentrated way so that a hole might be drilled in one part of an object while leaving the surrounding material cool and untouched. The color of a laser beam depended on its wavelength.

But the show would not be over. Mike would always bring along a small, low powered ruby laser. To the students' disappointment when he turned it on, there would be no fiery beam, no brilliant trail of light. "But if it's a ruby laser, isn't it supposed to be red?" someone would inevitably ask.

"It is red," Mike would contend. "The reason you do not see it is because there is nothing in the beam reflecting the light to your eyes so you can see it."

Then, he would turn the lights off in the classroom and strike a match, blow it out, and hold it carefully below the invisible beam. As the smoke would rise through the beam, the red of the ruby laser would sparkle in the darkness. The class would *ooh* and *ahh*, mesmerized while the visiting magician lit and extinguished more matches. Mike would revel in the thought that these young minds would one day grasp the real magic of lasers in ways their parents could not.

Though he would caution the young students to be careful around

On the Laser's Edge...30

lasers, he would not tell them about the secret side of lasers — the evil twin — that flourished behind the scenes: laser weapons, spawned in secret laboratories throughout the world, hidden and mysterious, locked inside a Pandora's Box.

Mike preferred to create good lasers, but his mind didn't always make the clear distinction of where the good ones ended and the bad ones started. It was up to other people to decide that. Even a bad laser, he reasoned, could be used for good if the people behind it had good intentions. Scientists in countries throughout the world were developing the warfare aspect of the science, and he felt that keeping our country safe meant keeping its lead. The U.S. could not afford to lose out to any country that might be a potential enemy.

With certainty, he loved all lasers, and he was always amazed at how easily the science of them came to him. When he met Maiman, he told him, "If it hadn't been for the invention of laser, I don't know what I would have been: probably a zero."

Mike was born Myron Clarence Muckerheide (pronounced Muck' er hide ee) on a mild, spring day in 1930. He was the oldest of four children born into the staunch Catholic home of Clarence and Esther Muckerheide in the city of Wausau: a thriving community along the Wisconsin River carved out of the pine forests in the 1800s by loggers, and the entrepreneurs who followed them. His younger brother died in his youth, leaving Mike and two sisters.

Although Mike's father had a good job at the local 3M factory, the Great Depression hit Central Wisconsin long and hard, and the Muckerheide family suffered under the crippling affects. "I can remember we really struggled," Mike reflected. "My Dad had a car and the car was up on blocks. He didn't have money to buy gas. He was a very hardworking man. He wouldn't take any help from anyone, no kind of assistance or welfare. He was a real German. He wouldn't do that. We really suffered during the Depression. We struggled terribly. We had a lot of patched clothing, and we ate a lot of rice and beans."

For extra income, his father electrified houses. Mike used to tag along. It was during these trips that he discovered his love for science and invention.

"A lot of homes still had gas lighting and we would run wires through the gas pipes; that's how conduit came to be. And I would help him. I'd pull the wires through the pipe and into the chandeliers and stoves," Mike said.

Electricity was still relatively new to the area, and misunderstood. One incident Mike remembered in detail. His father had electrified a large home belonging to a wealthy woman in town. One evening, after the job was complete, the woman paid a visit to the Muckerheide home. "She was really angry. She was livid, yelling at my father, 'You cheated me!'" Mike recalled. "She said that every time she turned on a switch, all the lights would go out."

After investigating the problem, Mike's father discovered that the woman had screwed fuses directly into several of the light sockets, creating a short. "Light bulbs had the same size sockets in those days as fuses," Mike explained. "And my dad told her, 'You can't do this. Either put a bulb in there, or leave it alone.' He asked her, 'Why did you do that? And she said, 'to keep the electric juice from running out.'"

Afterwards, his father took Mike into the basement and patiently explained the principles of electricity to his young son. "He explained to me the analogy of juice [electrical current] and water and they call it juice, but it isn't juice. And he told me that what this woman did was wrong and that this is a mistake that can be very serious. He explained all of this to me, and I was only seven years old. And that stuck with me from the very beginning."

From his first introduction, electricity fascinated Mike, and so did rockets. He and a neighborhood friend, Dicky Britten, used to visit the library to look up propulsion mixtures they could concoct from scratch. "We'd read about [Robert] Goddard and make mixtures to propel things," he said. "It used to drive my father crazy."

Mike was as artistic as he was scholarly. At home, he would spend hours making model airplanes and battleships from balsa wood and glue, trading rides in his homemade, toy Jeep in exchange for stickpins he needed for the construction. The ships he floated in a pool at nearby Marathon Park. He set the model planes on fire and sent them diving out an attic window to watch them crash and burn. When feeling more reserved, he would paint pictures on scraps of lumber with leftover house paint and shoe polish.

His fascination also included radios, and he built several of them while still in elementary school. "I made a lot of things then, but radios intrigued me," he said.

Not surprisingly, his favorite fictional characters were Flash Gordon and Buck Rogers.

On the Laser's Edge...32

One of the most life-changing episodes in Mike's early life was the death of his nine-year-old brother Clarence to polio. Not only did the ten-year-old Mike lose a close sibling playmate, but his quick illness and sudden death cast the normally reserved family into an unwelcome spotlight. The disease from which his brother died, Bulbar Poliomyelitis — also known as Infantile Paralysis — was the first for Wausau and it was highly contagious. Public notice, in heart-wrenching detail, was given in the local newspaper. The article noted that the Muckerheide family "was segregated by order of the department of health" and city and school nurses were preparing to examine all 600 students at St. Mary's School for symptoms of the dreaded disease, then monitored for a week. The health department also ordered "strictly private funeral services" for the morning following the young boy's death.[9]

If their grief wasn't enough to bear, the public stigma against the family was greater than what they endured as a result of their poverty. Back at school, Mike was shunned and ridiculed by his fearful classmates. It was the one defining moment that turned the sad and angry young boy inward. Books by the wagonload, checked out from the local library, became his companions, along with his tool shed creations.

He drew closer to his two sisters Mary and Carol: in particular the youngest, Carol. "I always looked up to Mike," she said. "As I grew up, I guess you could say I almost idolized him." Her eagerness to please him made her a willing guinea pig for test piloting his various contraptions: generally vehicles that lacked a braking system. "He built a Jeep that was as big as [a] couch out of scrap metal. He had fenders on it and a hood... He put me in it and sent me around the block, no brakes, so I couldn't do anything until I got back home." Another time he sent her down a hill in a brakeless cart. "I was always glad to do it for him," she admits of their exploits.

Mike was presented a coveted opportunity as a freshman in high school when he was awarded a scholarship to attend St. John's Prep School in Collegeville, Minnesota. Now as a young teenager, he would leave his home and his family for the first time. It was a prestigious move for the poor boy from Wisconsin, made possible

---

[9] "Heath Department Acts as Boy, 9, is Stricken," *Wausau Daily Record-Herald*, October 11, 1940.

through an uncle who was a priest with school connections.

Appearances were deceiving, however. The scholarship was not nearly enough to cover all of the expenses and Mike's family was too poor to make up the difference. "I had to work to put myself through," Mike recalled. "I remember my mother, the first job I ever had, she took me to the social security office and got this card and she told me, 'Now that you have this job, you're going to have to pay room and board.'" He wasn't even old enough to drive a car.

His first summer job was pulverizing scrap glass at a sash and door plant. He later had jobs working on a road crew, and blasting granite in a quarry. On one occasion at the quarry, Mike's crew didn't hear the blast warning and was showered with a rain of granite.

While a senior at St. John's Prep Mike would meet his future wife. Patricia "Pat" Kohls was a Minnesota native from Cologne who had accompanied her brother to the all-boys school for his enrollment. Mike was immediately taken by the soft-spoken, blond haired beauty and wrote her a letter. She wrote back from her all-girls school, Our Lady of Good Council, in Mankato 120 miles away, and a romance was kindled. They never seriously dated anyone else, and never regretted it.

By now, Mike had grown into a handsome young man standing 5' 9" with a shaved head of brown hair, and a preference for short-sleeved white shirts.

With no real direction in mind after graduating from St. John's, and still no money, Mike headed for the military. He attended U.S. Navy Hospital Corps School at Great Lakes, Illinois just as the Korean War was heating up. He spent four years, from 1950 to 1954, working as an operating room technician — essentially a scrub nurse — first at Great Lakes, then on the attack transport ship, the *APA President Jackson*, off the coast of Korea, and later on the ground inside Korea. As a medic, he routinely saw the horrible consequences of war: the mangled bodies of soldiers and civilians, which he helped patch back together again. He often thought: *Why would God put anyone in such a situation?* At times, even his own life was in danger. It was his daily letters to and from his high school sweetheart, (intercepted and censored first by the nuns at Mankato), that carried him through. Pat always asked if he was able to make church services that week. He said he only missed when duty called.

Upon Mike's honorable discharge from the Navy, five service

medals and scars from shrapnel wounds only hinted at his personal contributions. He rarely talked about his service to anyone.

Work as a medic, however, had been rewarding and it developed in Mike a desire to begin studies for medical school. He asked his father to help him financially, but the frugal, German patriarch gave his son a choice: either medical school or marriage. Feeling that a longer wait would be unfair to Pat, who was then out of high school and working as a secretary, Mike chose marriage. They were married in 1955.

To make a living, he continued working in operating rooms. It included the U.S. Veterans Hospital in Minneapolis where he once helped prep for the famous heart surgeon Dr. C. Walton Lillehei, the father of open-heart surgery.

Later, at St. Mary's Hospital at Rochester, he worked with Mayo Clinic doctors and high frequency equipment for heart bypass and brain surgery. "It was like an apprenticeship," Mike recalled. He would sometimes make suggestions for equipment improvements. "I just did it, and they liked it," he said.

Two years after their marriage, Mike and his bride headed back to his native state when he got a job at St. Mary's Hospital in Wausau. While still an operating room technician by profession, he had his own laboratory in the basement of his home where he worked on various projects. Among his early inventions was a blood pressure machine. "I tried to get it patented, but it was just too expensive," Mike said of the venture. "I finally decided I couldn't afford to proceed [with the patent process]."

His successes would eventually lead to more leniency at work, however. "I had a little room at [St. Mary's] where I kept supplies, and sometimes in the afternoon, they would let me go over there and work," he said.

In 1960, life changed dramatically and forever for Mike while watching the national news on television. It was the introduction of the laser by Maiman. Mike described the moment as "a quantum leap" for him. He was totally captivated and fascinated by what he saw: a tubular gadget whose bright, fantastic light seemed to leap into reality from the comic books of his youth.

The young lab technician set to work to build his own laser. In his home basement, with few supplies available to him, he read, tinkered, and experimented. A former neighbor and family friend commented that she knew when the basement windows of the Muckerheide home glowed with strange colors in the night that "the mad scientist was at work."[10]

Because laser science was still so new, a lot of what Mike learned was through his own experimentation. His first laser wasn't very good, he admits. It was a small diode laser pumped with a strobe. Most of the components were made from junk. "My wife always said that I would never have been able to make a laser if it hadn't been for the junkyard," he said. It was all he could afford, and all he had access to using.

Simultaneously, Mike was working on another invention that would move him into a small burst of the worldwide spotlight and move him rapidly closer to full-time research. It had nothing to do with lasers, but with the electrocardiogram (EKG). He developed a system to send EKG signals over telephone lines.

At the time, the small, isolated hospitals in Northern Wisconsin had no cardiologists on staff, so EKGs were being sent by mail and Greyhound bus to specialists in Wausau. It was a time-consuming process that cost the lives of many critically ill heart patients. Mike believed that he could change the signal to sound, send the sound by phone and demodulate it on the other end, turning it back into an EKG signal. He had successfully tested his hypothesis from room to room, but needed the support of a telephone company to test it over long distances. He decided to approach the General Telephone Company office in Wausau. The trick was getting around the secretary. "I kept going over to GT to convince these guys to let me use a line to try this," he recalled, "I kept trying to see this one guy, Robert Glompski."

His dogged persistence obviously annoyed the man's secretary. "She kept telling me she didn't want to see me around there anymore. That's when I learned that the first word of *secretary* is *secret*."

Finally, Mike decided on a way around the secretary: he would stalk Glompski. He would wait in the General Telephone Company parking lot for Glompski to leave his office for the evening, jump out of his car, and command his audience. It paid off the first evening. Glompski listened to a quick description of Mike's EKG plan, and

10 Interview with Janet Volpe.

On the Laser's Edge...36

invited him back to his office to discuss it further. He liked what he heard, and decided to give Mike the chance he wanted.

On May 18, 1963 their first phone-transmitted EKG traveled 20,000 miles from Wausau, Wisconsin to Honolulu, Hawaii, and back to Wausau in less than a second. Both a human heartbeat and the heartbeat of a white mouse were sent to prove that the transmissions would be successful no matter how faint or rapid the heartbeat. Though stopping short of claiming to be the first (something no one was willing to take the expense to prove) the success of the experiment put Mike on the leading edge of such technology and Mike began getting phone calls from physicians and scientists around the country.

In October of 1965 Mike would repeat the experiment with the help of local television newsman Howard Gernetzke to prove its ability to reach longer distances. Gernetzke took a recording of an EKG to Bangkok, Thailand and successfully phoned it back to Wausau. The transmission, like the one to Honolulu, had only one-tenth of a percent error rate.

The result was revolutionary for small town practitioners in hospitals throughout the Northwoods: places like Park Falls, Antigo and Tomahawk. They were able to do in seconds what had taken them hours, sometimes days, to accomplish. By getting the EKGs to cardiologists sooner, they were able to diagnose and treat patients more effectively.

Also in the mid-1960s, General Telephone started using lasers, namely for the transmission of a program they sponsored on television, *The Gary Moore Show*. Mike wanted to try EKG via laser, so General Telephone gave Mike his first commercially manufactured laser for experiments. He conducted a successful experiment on May 11, 1964, which transmitted Robert Glompski's heartbeat over a 30-foot beam of light from Madison, Wisconsin to Wausau, a distance of about 145 miles.

By now Mike was working for Wausau Medical Clinic who had lured him away from the St. Mary's Hospital with promises of better pay and his own research laboratory. He was also being paid a small commission for every EKG transmission. Once again, however, lack of financial backing prevented him from patenting the system.

The move to Wausau Medical Clinic proved beneficial for Mike in many ways, not the least because of a man he met there. Dr. Ronald

Uecker was a radiologist who shared his interest in discovery. Like Mike, Dr. Uecker was brilliant, creative and a Wausau native who liked to invent. He directed the hospital's radiology department and at the time was working on several of his own patents, some of them related to his medical work and others to his passion for sailing.

Dr. Uecker was one of the first to recognize Mike's genius. The two men would sit and talk for hours both socially and professionally about research. Of course, Mike's favorite topic was lasers. On visits to the Uecker home he would regale his listeners with promises of things to come: laser hydrogenated cars, laser guided missiles, laser surgery. "It's just like Buck Rogers," he would say enthusiastically.

The average person, even the average doctor, found Mike's promises ludicrous. It was, after all, the 1960s when the heralded promises of lasers were yet to be delivered. Fortunately, Dr. Uecker provided Mike with something he needed: someone who really believed in him and who could open doors in the medical community generally left closed to someone lacking academic degrees and credentials. Together they would fight skepticism with success.

One of their first collaborations was comparable to the EKG experiment: the development of a system to transmit x-rays over the telephone line. They were successful in February 1968. As with the EKG, the transmissions were believed to be the first ever; yet, the news resulted in a mere mention in a technical manual. Still plagued by inadequate funding to develop the system further, the project was scrapped, becoming nothing more than a footnote in medical history. It did, however, solidify the team of Uecker and Muckerheide and they would continue to collaborate for several more years to come.

Back at General Telephone, however, things were beginning to sour. According to Mike, a GT executive, climbing his way to the top, abruptly halted all local EKG transmissions. The executive was helping GT formulate its own plans to send EKGs by wire, effectively cutting Mike and Wausau Medical Clinic out of the picture.

Knowing that patients in the Northwoods would die unnecessarily without the quick transmissions, Mike fought back. "I told them, 'if you shut me down these people up north are going to be without EKG. They can't send it in for heart analysis. They don't have doctors that are cardiologists. You're going to be killing people.'"

Mike contacted his congressman, David Obey, who in turn contacted the Public Service Commission. The Commission was successful in having the transmissions continued. By now, however, it was too late to mend fences. The exclusive partnership between General Telephone Company and Wausau Medical Clinic collapsed.

Though it dampened Mike's spirits, his head continued to spin with new ideas. If his brief and exhilarating experience with General Telephone Company had done nothing more, it had provided him with a real, working laser. He latched onto the technology with exuberance and incredible ease." I don't know where it comes from," he said. "It's just in my head. It has always been so easy for me."

Though the "genius" label was often given to Mike, it came with restrictions. "There's so many highly trained people...that know more about the theory of lasers than Mike, but in terms of how to make one work, how to build one, how to modify something to work better, Mike was as good as any..." said Darrell Seeley, who first met Mike while a physics professor at the Milwaukee School of Engineering. "What impressed me was his ability to think things through rather than theoretically calculate or design that way...He was very creative and he did an incredible amount with his limited formal training. That was what was unique about Mike, and his enthusiasm. His drive."

Mike's colleagues often wondered how much more he might have been able to contribute if only he had enjoyed a greater academic background; if he could have avoided lost time in the pursuit of basic learning that led to the obvious dead ends. At the same time, they realized his naivete sometimes led him to discover things that others missed in their textbook approach. "He didn't know what he wasn't supposed to do," as one acquaintance put it, "and sometimes that paid off."

The timing of laser's development and Mike's developing career could not have been more perfectly timed. Like the self-taught computer nerds of the 1980s and '90s, Mike was on the ground floor when the foundation of laser science was being laid. It was a new and misunderstood science that a new and misunderstood scientist could latch onto. He not only learned to navigate his way through it, he learned how to manipulate its beams into new directions of discovery.

While laser knowledge was easy for Mike, the politics of it would nearly become his undoing. Behind the science, lasers were being developed with the capacities to save lives in incredible ways, and to destroy them just as incredibly: the right ingredients of power coveted equally by men of good, and men of evil. It was a hard lesson Mike would subsequently learn.

CHAPTER FIVE

The most significant domestic terrorism threat over the next five years will continue to be the "lone wolf" terrorist.
— *Federal Bureau of Investigation Strategic Plan 2004-2009*

As soon as the laser was introduced in 1960, the scientific world sat up and took notice. From that point on, in laboratories around the globe, lasers were being raced into evolution at the hands of scientists. Medical researchers saw the potential of healing and life, while military developers saw that of destruction and death.

U.S. publications reported that by the late 1960s, the USSR was testing the use of laser in nuclear fusion for hunter-killer satellite missions. Not to be outdone, the U.S. was rapidly and secretly gaining its own foothold. By the 1970s, military experts knew that lasers could and likely would revolutionize warfare, and occasional mishaps gave rise to suspicions that some laser weapons were already being used — tested — in combat and surveillance situations. Mainstream publications were starting to report this; yet, it was a science that remained shrouded in mystery and secrecy. The public at large knew nothing about it, and the science was so far beyond average comprehension that most people wouldn't have felt much of a connection to it anyway. Weapons were just weapons, just another way of destroying things.

Perhaps compounding the popular disbelief was a weary, public attitude towards war. The development of more weapons was the last thing the average American was in the mood for.

Al Iwen, however, represented an element of society still very much in the mood for war. At a time when entire countries were scrambling to develop laser weapons, he set in motion a plan to create his own. Armed with educational credentials from Indiana Institute of Technology, Beloit College, the University of North Dakota and the University of Chicago, and a work resume listing the Saturn 5 program, and biological satellite studies, he might have been labeled a good visionary, if only he'd stuck to better intentions.

As it was, throughout the remainder of 1976, Iwen devoted considerable time and energy to planning his own laser war. It would take months for Mike to understand what he really did have in mind. They started meeting sporadically, every few weeks. Sometimes Iwen would talk a lot about his plan, hunkered over airplane schematics and aeronautical charts. At other times, he would mention nothing at all.

Mike claimed that Iwen had "an odd way about him." More specifically, "he was very secretive." Mike would take his cue from him, offering little unless asked, keeping his attention keyed to discover what he could about the recalcitrant chemist.

It didn't take long to realize that Iwen moved in many circles, keeping references about one circle of friends vague to the next. His laser circle remained small: just Claussen and Mike. Claussen, however, didn't seem to have any obvious role to play; he was just someone Iwen trusted — a friend — and the original link to Mike. He was the organizer, setting meetings and keeping everybody informed.

When their meetings included members of the Posse, Mike noticed that Iwen was careful never to discuss the laser plan. He finally concluded that the Posse as a group wasn't directly involved: Iwen's plans were his own. "Iwen thinks he's bigger than the Posse," Mike told Southworth. And they agreed that Iwen was using them to get what he wanted.

Today, Iwen would be labeled a "lone wolf": a popular buzz word in terrorism.

The FBI has a theory about the lone wolf; he likes to operate alone because a larger group can be a liability: too many people talking about too many things can kill a plan. He has chosen a solitary path; yet, may need the help of others to ultimately carry out his mission. Because a lone wolf leaves fewer tracks, he's harder to hunt. The FBI thinks it is the lone wolf that currently poses the greatest security threat because of the Bureau's success at keeping tabs on more troublesome groups.

The term "lone wolf", however, didn't get much play until the turn of the 21st century when the FBI was worried about Y2K pranks. Once again, Iwen was ahead of his time: a lone wolf long before it was popular.

Mike suspected that Iwen was selectively drawing people into his plan, one by one, from his inner circle of friends. Rarely, however, would a name surface. Instead, he would use generalities, claiming the allegiance of an airplane pilot he knew, or some anonymous retired colonel, or another scientist. Carefully he would dole out information only when necessary. By intention, he was the only one who knew the full scope of the plan or the identities of all the players.

Although some of the players were Posse members, Iwen's plan was more ambitious. The Posse was devoting most of its energies to gaining local control and were largely reactive. The group Iwen was gathering around him was set on activism on a more worldly stage, in a big and aggressive way.

It's not surprising that Iwen's group was small and tightly knit. Few people would have known enough about lasers to accept his ideas as viable, and keeping the lid on such a grandiose conspiracy would have been difficult. More importantly, such a serious conspiracy as he was hatching would have been doomed if he had divulged the details haphazardly. He needed to handpick his members, and their choosing would be based on what skills they could bring to the plan. It would take another year of trust building before Mike could himself see the larger picture painted in Iwen's mind.

While Iwen remained vague about his intentions, something happened to steel his resolve and set a fire beneath them. It started as a real fire that severely damaged his laboratory in Mosinee in the fall of 1976. By now the U.S. Food and Drug Administration was raising concerns about his laetrile operation. Feeling the heat in more ways than one, the owners of the business moved east approximately 135 miles to a former dairy building in Manitowoc where it was renamed U.S. Pharmaceutical, Inc. On April 5, 1977 officials arrived at the Manitowoc lab to conduct an inspection. Iwen turned them away. When they came back a week later, they came armed with a federal search warrant and the assistance of local police and U.S. Marshals. They seized items they believed were being used in the manufacturing of laetrile, and the plant was ordered closed.[11]

At first, preoccupied with his legal entanglements, the number of

---

[11] By the late 1970s, laetrile had turned into a hot, political issue. The War on Cancer had been declared in an act of Congress in 1971 and no cure had yet been found. People were desperate and willing to accept the hope of an unproved drug. Legal action was sometimes slow and indefinite as the public, politicians and FDA sorted out the issue.

On the Laser's Edge...42

meetings instigated by Iwen declined. When they started again, they increased to a degree that it was obvious he had not stopped thinking about lasers or laetrile. In fact, lasers had moved to front and center of his thoughts.

Now, the three men began meeting every couple of weeks. Iwen's paranoia grew, and along with it the landscape of his laser plan.

If he were operating today, he could have bypassed Mike and popped in at one of the world's blackmarkets for lasers and purchased a finished produced, or paid-off one of the many thousands of laser experts now in abundance. As it was, a person couldn't find a laser scientist just anywhere back then. Whether he wanted to or not, Iwen had to trust Mike if he was going to make his plan work.

Mike was not thrilled at becoming Iwen's confidant. He kept thinking, 'any day now, he'll be arrested and taken out of circulation.' But, nothing happened as the days turned into weeks and the weeks into months. As he waited, Mike continued to help the FBI gather more information.

Mike's reports to Agent Southworth now included the names of specific properties Iwen wanted to target in his private laser war. Included was an office building at the University of Chicago, where Iwen had spent some research time, and various synagogues in New York and Wisconsin. The Chicago building, at 1313 E. 60th St., was one commonly criticized by anti-Zionist groups because they believed it housed several agencies of the Rockefeller Foundation and was being used as the headquarters for a secret government.

On the world front, Mike said Iwen wanted to, "...fly against Cuba and attack Cuba, and by attacking Cuba produce a confrontation between the Soviets and the United States so that there would be a Soviet retaliatory strike against the United States."[12] He would push the blame onto the CIA, an agency he strongly disliked, and find refuge in a foreign country. His chosen weapon would be a powerful, carbon dioxide laser mounted in the nose of an airplane.

While the plan sounded far-fetched, Mike knew the scientific possibilities. Not only that, he had access to the right parts and equipment to construct the laser. Mike routinely worked with carbon dioxide lasers — the most powerful to date. "I could do this very easily," he kept stressing to Southworth. "And I could make twenty instead of one."

12 Grand Jury testimony of Mike Muckerheide, May 2, 1979.

As Iwen became more determined, it posed a problem for Mike: either he had to start showing some signs of progress, or find a way to get out of his predicament all together. He hadn't done much besides offer some general information and discuss airplane schematics and that wasn't going to carry him much farther. Although he really wanted to bail out, he knew it would be difficult without causing suspicion. As well, the FBI was reluctant to lose its inside connection and kept nudging him on. As impossible as Iwen's plan may have sounded, the Bureau had a history of dealing with the absurd. They dared not turn their backs on Iwen until discovering how serious he really was, especially given his resume.

One huge problem presented itself to Mike if he planned to continue on, however. While he was able to limit the details he offered Iwen about lasers — because a lot was still unknown about the science — Iwen's scientific background and basic understanding of lasers made it hard to pass off bogus information. That left Mike only one effective way to move Iwen's project forward while pushing it backward: it came down to money.

During the early years of development, lasers were very expensive; not because of any special, expensive materials needed, so much as the cost of developing the technology and the simple principles of supply and demand. At the time, lasers cost the equivalent of $1,000 a watt. In essence, a 50-watt laser cost $50,000, but it was the technology that cost money, and Mike had the technology to make that same 50-watt laser for just a hundred dollars. He wasn't about to tell Iwen that, however. Instead, he kept telling him how expensive it would be, and his estimates escalated as time progressed.

Part of what helped make this delaying tactic work was Iwen's changing plans. As he became more specific, he would add and subtract details. As it changed, parts of it became more feasible than others, but each part added to the complexity and overall cost. At least, it was a good argument, and one that Mike could use to stall progress.

It was a good tactic since Iwen's laetrile operation, once rather lucrative, had fallen on hard times and his legal bills were eating away at the profits. Not to be beaten, however, he quickly came up with an answer. He was certain he could find a foreign country interested in buying some laser technology.[13] Once in, he would set up a

---

13 Sworn statement of John Claussen.

laetrile operation to keep generating money for Mike's work. Latin America seemed logical because several countries there were in civil turmoil and anxious for weapons. It was also relatively close for the numerous trips it would require, and it offered a favorable climate for the growing of apricots: the primary ingredient in laetrile.

One might wonder what might compel someone to want to start a laser war. Even a look at Iwen's background doesn't adequately answer the question.

He had graduated from Merrill High School, 15 miles north of Wausau, in 1949. He was raised on a small dairy farm, and wasn't very active in school, listing only forestry in the annual high school yearbook.

He went on to earn a Bachelor of Science degree from Indiana Institute of Technology in 1954 and did further undergraduate and graduate work in psychology and chemistry at Beloit College at Beloit, Wisconsin, and the University of North Dakota. He did research work at the University of Chicago.

His education and employment were interrupted from 1954-58 when he served in the U.S. Air Force as a special weapons fusing systems technician.

He entered the space program in 1965, working with Boeing Aircraft Corporation on rocket launching for the Saturn 5 program, moving on to Brown Engineering Company at Huntsville, Alabama to work on the space shuttle program in biological satellite studies. From there, in approximately 1969, he taught chemistry at the University of Southern Illinois for about a year-and-a-half while working towards his doctorate degree in chemistry. [14]

Despite an auspicious beginning, by 1972 Iwen was back in north central Wisconsin running for mayor of his hometown. News articles about his candidacy listed him as "an unemployed aerospace engineer, currently cutting pulpwood."

Interestingly, when Agent Southworth attempted to verify his work record, he hit a wall. "Every time I attempted to determine the validity of his identity, bells went off," he recalled. "I can remember

---

[14] Career information obtained from Grand Jury testimony given by Albert C. Iwen, April 13, 1977 for case #77-CR-47, *United States versus Albert C. Iwen and Douglas Evers*, U.S. District Court, Eastern District of Wisconsin; and from the *Merrill Daily Herald*, January 10, 1972.

sending the material in, doing inquiries, sending out leads to verify who he was and basically being told from Washington that I didn't need to do that because 'you're going to mess something up.'"

Upon his return to Wisconsin, still a bachelor and now in his 40s, Iwen became almost immediately involved in ultra conservative movements, in particular the American Party. They were attempting to set up chapters in every county and Iwen was leading the effort in Lincoln County.

When Iwen ran for mayor in the spring of 1972, he laid out his agenda for change. "My primary aim is to counteract the accelerating trend toward reducing the authority of our locally elected officials," he said in his announcement. "The current unconstitutional programs for phasing us into a dictatorial regional metro complex, wherein we are to be ruled by a self-appointed intellectual elite cadre of experts 'who know what is best for us,' must be resisted and reversed."

He failed to win the election, so ran for the 12th State Senatorial seat that fall in a statewide push by the American Party to gain state representation. Again, he lost. At various other times, he ran unsuccessfully for the Merrill City Council and the local school board.

His election numbers were never very good. People saw him as a strange fellow squandering a good education on foolish ideas.

CHAPTER SIX

Don't fall before you're pushed. – English Proverb

Mike was beginning to feel uneasy about the FBI. Agent Jerry Southworth was leaving the Wausau Resident Agency, having taken a position at Headquarters in Washington, D.C. The move had been prompted by Southworth's infiltration of the Posse Comitatus and fears for his own safety. He had brazenly started attending Posse meetings to see firsthand how they operated, pluckily admitting his employment. Some of the things the Catholic-raised agent saw and heard made his hair stand on end: about ten members would show up at each meeting decked out in rifles and handguns, their conversations peppered with hatred towards Jews, Catholics and racial minorities. One of the bizarre stories circulating was that the Catholics were kidnapping and killing non-Catholic babies and burying them on the grounds of a local convent. The stories were unbelievable, but equally so was the realization that people believed them.

Some of the Posse members embraced Southworth's presence thinking they had won a convert. Others believed he was obviously spying. Southworth was no convert, and when his attendance at the meetings began to cause dissension in the ranks, he started looking for greener pastures. "They were a scary group, and they knew where I lived, and I had a family," Southworth said. When the Bureau offered him a promotion to Washington, he packed up and left.

Knowing how closely Iwen brushed shoulders with Posse members, Southworth left Mike with one piece of advice: "These people are carrying guns. Don't be going anywhere with any of them." It was advice that Mike didn't always have an option to heed.

Southworth's departure was a blow to Mike. Being an informant required a great deal of trust on everyone's part, and the one person Mike had learned to trust was gone. He knew that the story unfolding was too unbelievable for most people to comprehend, and he knew that it had taken Agent Southworth time to believe in him. A second agent assigned briefly to the Wausau office during the Southworth years never did. Now it required starting over with somebody new.

It couldn't have happened at a worse time. The tables of investigation had turned after the death of the FBI's first director, the indomitable J. Edgar Hoover, in 1972 after 48 years at the helm. Like vultures attracted to a fresh kill, congressional committees swooped in to investigate several suspected abuses. The agency's image was left damaged. The final bone was picked by Attorney General Edward Levi in 1976 when he announced his decision to severely restrict the way the FBI could investigate organizations. Levi was a big supporter of the right to free speech or expression and was more than a bit perturbed by what went on during Hoover's final years at the Bureau when agents infiltrated Vietnam War protest groups, civil rights organizations and the women's liberation movement.

Levi's new curbs went into effect just as the laser case was gearing up. What affect they might have on Mike's situation had him edgy.

More recently, Nixon's Watergate fiasco had pulled Hoover's interim replacement, L. Patrick Gray, into the controversy. Under pressure, Gray withdrew his name from consideration as the Bureau's permanent director, leaving a temporary void. The incident left many Americans wondering what role the FBI might have played in the Watergate episode, and it left Mike wondering who was really in control of the FBI; what would happen to the information he had provided them?

To protect his family, Mike shared few of his experiences or fears with anyone, not even to Pat. It was a natural reaction for a man who, since the age of 10, had become emotionally isolated and had taken upon himself the Herculean role as the family protector. More significantly, Pat had been diagnosed with breast cancer soon after the birth of their daughter in the late 1960s and had suffered through a series of treatments and mastectomies. Mike did not want to burden her with his battles with crazy extremists when she had her own battles to fight between life and death. It was a heavy, psychological burden to carry alone. Sometimes Mike worried about the case, and about his own sanity. That isolation would only sink him deeper into the mire of the conspiracy.

Southworth's replacement was Jack Page. In some ways he was too much like Mike: serious, hard-nosed and tight lipped. Page's

On the Laser's Edge...48

excuse was that he was from the taciturn state of Oklahoma. Time spent as a pilot in the Vietnam War was likely a contributing factor. As a result, the agent and the scientist would take some time adjusting to one another; their first encounters danced out like two cautious animals meeting for the first time.

No different than Agent Southworth, Page would venture slowly into the case, testing Mike and the information he delivered. One time, Page asked Mike to meet him. At their rendezvous site, the agent instructed Mike to get into the back seat of his Bureau car. There was another man riding in the front seat, but Page never introduced him, and the stranger remained quiet, looking straight ahead. They drove away and, as they headed down Grand Avenue in front of Wausau Hospital, Page asked Mike bluntly, "Could you make an atomic bomb?"

The question startled Mike, but it had merit.[15] Scientists were exploring ways to use laser to ignite the explosion of an atomic bomb. "Well," he told him, "I don't know if I could get it to go off."

Agent Page immediately turned the car around and dropped Mike back at their meeting site. He never brought the subject up again in conversation, not even to explain why he'd asked.

It bothered Mike that Page was like that: letting things go by without explanation.

Tom Burg once said, "Mike loved his family, but his life was

---

[15] From an interview with Mike's first photographer, Ray Gehring: "I don't remember what laser [Mike] had when the people came down from Minneapolis Honeywell. [Mike] had a laser down in his basement and these guys did not want him to energize it – to fire this thing up – while they were inside the house. Mike said, 'this one is capable of creating a nuclear event.'...but he didn't fire it up because these guys said, 'you could start a chain reaction and take Wausau off the map. We're out of here.' They said they would come back some other time. They were afraid of the thing. Apparently, just looking at it, they knew it had some potential." Gehring isn't sure what frightened the Honeywell men, but Mike was not adverse to dangerous experiments. On one occasion, Gehring was present when Mike was experimenting with a small amount of Uranium-235, the active element in nuclear fission. "He had this little, brass, vacuum chamber... and there was a glass top on it so we could view it, and he was bombarding this U-235 with his laser, and there was like a helix coming out and through. We never saw this at the time, but on the pictures we saw a trail of light coming off, and Mike said, 'Oh, my gosh! Look at what you caught! Look at what you caught!"

Code Word Tikal...49

lasers." Mike's family said it was true. When others failed him, he still had his lasers and he clung to them like a man overboard left clinging to a raft.

At the same time as the conspiracy began to suck him under, Mike's career was beginning to take off. He was establishing a good reputation outside the Wausau area for his expertise in lasers and his local work was starting to settle in after a rough start.

His projects were numerous and varied, involving experiments that would help advance the sciences of tumor detection, heart surgery, rheumatoid arthritis treatment and birthmark removal.[16] He was beginning to travel around the country to confer with top medical researchers.

Just as his interest encompassed all aspects of laser, so did Mike's research. In addition to medical lasers, he was independently exploring the development of surveillance systems, large, exterior lighting systems, and riot control devices.[17]

He was also designing laser labs and bringing them to full operational status. He particularly developed strong ties to the University of Wisconsin - Stevens Point and, soon thereafter, North Central Technical Institute (later renamed Northcentral Technical College) in Wausau.

Pursuing separate opportunities for a time, Mike and Dr. Uecker teamed up again in the early 1970s, this time with researchers at

---

[16] Interview with Mike's first photographer, Ray Gehring: "One time when we were at Stevens Point, we were there with a bunch of people standing around, and their clothes were strange. So I learned they were from Russia. Some of them were bodyguards and one of them, according to Mike, was a member of the Politburo. He was there because his daughter had a large birthmark and they came to have it removed. If it was successful, I don't know."

[17] Also from Ray Gehring: "One time, we were down at the airport in Wausau. Mike was there, and all kinds of uniform guys were there, and he had this laser weapon at this time. This was possibly at the beginning of Vietnam, because I was just out of service... he had this device, a laser weapon... and when he was at the airport there were some representatives from, I think, the Los Angeles SWAT Team and the Chicago SWAT Team. They were interested in buying this thing... he said to me, 'hey, what should I do with this? I'm between two positions. I can either put this into production, or sell it.' He was also being harassed by a bunch of the banana republics. They wanted to buy this."

On the Laser's Edge...50

Battelle Memorial Institute in Columbus, Ohio, headed by physicist Dr. Philip Mallozzi. They worked on a procedure that would one day be patented: a method for applying radiation at a desired location using laser-generated plasmas as a source of x-rays. The intended result was easier, faster diagnosis of tumors: a project dear to Mike's heart because of Pat's struggles with cancer.[18]

While his uncanny ability with lasers kept him working, it also made him a target for criticism even among colleagues. As with many gifted researchers, his vision and comprehension reached beyond ordinary understanding and he didn't always temper predictions for his audience. In his case the revolutionary science of lasers only helped to exaggerate perceptions that he was either crazy or, at best, eccentric.

"It was bizarre stuff," James Lombardo conceded of some of Mike's scientific contentions. Lombardo was an audiologist who enjoyed spending time with Mike trying to develop holograms of the inner ear as a diagnostic tool, and exploring the effects of noise on communication. It was not unlike Mike to stop in Lombardo's office for a brief conversation on "black holes and antimatter matter," leaving his listener dumfounded. "I'd think, what the hell is he talking about?" the audiologist remembers. When it was time to go to work, Mike was always eager to "snap the atmosphere" for his young lab partner.

"I never doubted that Mike was a brilliant, self-made genius," Lombardo said. "There's no question in my mind. And I enjoyed his mind. I enjoyed the inquisitiveness of it."

Lombardo shared a glimpse into Mike's inquisitive mind in a story about one of their lighter moments together in the hospital's laboratory. The two men were working with an audiometer: an instrument used to test the ear's ability to detect sounds over a range of fre-

---

[18] For the most part, Mike's work was very practical in nature and had a connection to his routine work, such as the case of his cancer research. Mike's first photographer, Ray Gehring, recalled some of his work in this area: "I remember he had a piece of what looked like a cube of Lucite and it was supposed to be to the laser the same as body tissue. Within this cube he took some tissue, some kind of cancer tissue, and imbedded that into the Lucite... he blasted it with the laser and the tissue was gone. Inside that little cavity there was no residue, no moisture, no ash. It was gone. He thought at that point that he was onto something with cancer treatment."

quencies and intensities. At one point, Mike took the headset off the audiometer, exposing the diaphragm. He placed the diaphragm on a piece of Plexiglas, dumped a fine powder on the Plexiglas and then turned on his neodymium laser so the light would hit it. Reflected on the ceiling were the results: odd, holographic patterns of light. Lombardo manipulated the audiometer controls to change the kaleidoscopic patterns. "I remember one night we sat there and just looked up at this for hours," Lombardo recalls. "And Mike would just say, 'I wonder what we're seeing? What the hell are we seeing up there?' I'm sure we were just seeing patterns of vibration changed with frequency and intensity, but I don't think anyone had seen it as clearly as that. It was interesting. Very interesting."

Not everyone found Mike's extemporaneous experiments and conversations so intriguing. There were certainly doctors, generally those with no practical experience in research, who believed that everything Mike said was a figment of his imagination. They believed that either he was a genius or a huckster who was fooling everyone. It didn't help that he did not have a college education and that, as a result, he sometimes bumped against a ceiling of scientific reality that was so obvious to others. Of his lack of a college degree one doctor commented that Mike did not deserve to wear a suit and tie to work. Mike was quick to anger and slow to forgive and the criticism only fueled his determination to prove them wrong.

The year 1975 was not a good one for Mike. Frustrations dealt by his professional detractors were multiplied by the sudden death of Dr. Uecker, one of his most ardent supporters. Uecker died suddenly of a heart attack while attending a conference in Chicago. Some people thought the death was suspicious because the slender Dr. Uecker was only 42 years old and in apparent good health. In reality he was under tremendous pressures in the years leading up to his death. Long days, minimal sleep and the pressures of research politics had proved a deadly combination for the overworked radiologist/ researcher. Mike was devastated and angry. "They killed him," he said bitterly, referring to Uecker's own detractors. He meant figuratively, but not everybody took it that way.

By now, Mike had also tired of the cynicism and the restraints placed on his research at Wausau Clinic. In an act that was anything but pretty, he left; his departure mutually appreciated by his superiors who were weary of a hired hand who was not afraid to speak his mind.

On the Laser's Edge...52

Mike moved on to become Research Director of the A. Ward Ford Memorial Institute headquartered in Wausau. The Institute was established by Caroline Mark and her engineer husband, William B. Mark. Caroline was an heir to the IBM fortune through her grandfather, A. Ward Ford. Funding for at least part of Mike's laser work at the Institute came from his greatest supporter, William "Bill" Mark, who established the nonprofit C-Corporation, Mark Laser Systems, to funnel the research dollars. Some of Mike's early work in the field of cardiac research was made through the collaboration of the Institute and Mark Lasers, including his first medical-related laser patent submitted in September 1976; during the early stages of Mike's contact with Iwen. As with most of his unclassified patents, this laser had a medical application: a system and method of diagnosing conditions in inaccessible areas.

At about the same time, a visiting cardiologist, Dr. Ellet Drake, introduced Mike to researchers at the Henry Ford Hospital Research Center in Detroit where Drake headed the cardiology department. This began Mike's long association with the hospital on heart-related research.

Another significant colleague in heart research during this era was Dr. Mahmood Mihroseini whose own work in myocardial revascularization by laser (bypass surgery) was world renown.

Drake also introduced Mike to the man who would be called the Father of Laser Surgery, Leon Goldman, M.D, a dermatologist at the University of Cincinnati.[19]

By 1977, Mike's professional life was beginning to settle. He had found a comfortable home at the A. Ward Ford Institute and Mark Lasers. The world of lasers was opening up, and he was finding a more welcoming response to his work outside his hometown. He became protective of that growing success, allowing no one, not even those who worked with him night and day, to ever suspect what terri-

---

19 Mike would contribute a chapter in Goldman's book, *The Biomedical Laser: Technology and Clinical Applications* edited by Dr. Leon Goldman, published by Springer-Verlag, 1981. The chapter was entitled "Laser Medical Technology for the Twenty-first Century." In 1980, Dr. Leon Goldman and Dr. Ellet Drake established the nonprofit American Society for Laser Medicine and Surgery, Inc. using seed money from the A. Ward Ford Institute. It presently has about 3,000 members and is headquartered at Wausau, Wisconsin.

ble, dark shadow followed him.

That shadow moved perilously close to exposure at times despite Mike's best efforts. James Lombardo was a case in point. Sometime in 1976 the audiologist, a runner who enjoyed competing in marathons, received a phone call from a man professing to be a runner who was looking for someone with whom he could train. He had been connected through a mutual acquaintance.

The stranger who showed up was a rather unusual character for the sport: a man of about 30, with a definite waddle to his gait. Over the course of the next year they occasionally ran together, sometimes in the company of their mutual acquaintance and sometimes not.

The first time the two ran together alone, the man arrived with a hunting knife dangling from his belt. "What the hell's that for?" Lombardo asked him. He replied that he brought it along to protect them against the packs of wild dogs that roamed the neighborhood. It seemed strange to Lombardo, but more so in hindsight, as did their frequent conversations. "So many times when we ran together the discussion would always come around to what research I was doing with Mike Muckerheide," Lombardo said. The questions didn't seem particularly out-of-the-ordinary at the time; after all, the audiologist was also fascinated by the work of the laser scientist whose work had been exposed in local newspaper articles.

Eventually, however, Lombardo became annoyed by the company of someone who obviously had no real aptitude for the sport. After competing together in a local marathon, they went their separate ways, Lombardo never suspecting that the inquisitive man, John Claussen, was involved with Mike in a strange conspiracy.

If Mike knew about Lombardo's association with Claussen, he never said a word. By the spring of 1977 he had his hands full with everything else he was attempting to juggle, and now his professional life was taking a detour — one of many — because of the laser conspiracy.

CHAPTER SEVEN

Courage is the art of being the only one who knows you're scared to death.
— Harold Wilson

Mike was convinced that he was going to die. The two men who would kill him were escorting him along a path on a West Virginia hillside. Off to one side was a pile of dirt that he knew was a freshly dug hole, and he believed that that was where they were going to bury him.

It was the fall of 1977: a great year for Mike's career, a terrible year for the laser case. This was his third trip out-of-state pursuing Iwen's wild dreams, and it promised to be his last.

Flash back to April: a year to the month since his introduction to the creative chemist. Their conversations had taken some unusual twists and it was obvious now that Iwen had become tired of talking about his plans and was ready to make them reality.

His first step was to enlist the assistance of a former colleague to supply the location for a secret laboratory where Mike could work to develop laser weapons. He arranged for Mike to journey to the remote hill country of West Virginia to check out a possible site at the home of John Hartl.

Hartl and Iwen had met while working at Brown Engineering Company in Huntsville, Alabama in the 1960s when the U.S. Army was testing lasers at the nearby Redstone Arsenal.

Iwen, an amateur pilot, had planned to fly Mike to West Virginia in his Comanche airplane. The FDA's visits to his Manitowoc lab just days before their scheduled departure, however, had thrown him off balance. Instead, he purchased a ticket on a commercial flight for Mike to make the trip alone.

Mike and Hartl had never met, so Iwen provided a letter of introduction. It wasn't a normal introduction, unless one is used to dealing with neo-Nazis. It read:

Dear John,
 I haven't had a chance to call you as of this moment to explain. Hope to before this is delivered.
 Due to legal complications, it would be unwise for me to attempt to travel out of state until the problem is settled.
 The bearer of this comes to you by direct authorization of DF 5th R. He is "V" chief VR. Description 5'9", blue eyed, slightly balding.
 His military discharge card of Sept. 1954 is signed by E.G. Connelly, LTUG, USNR.
 Regret not being able to come personally but am afraid to explain, for this would reveal our established identities if this note should fall into enemy hands, unlikely though this may be.
 V=vorshuchungs=research

As ever
DF

The initials "DF" referred to Iwen's code name: der Führer of the Fifth Reich. He referred to Mike in the letter as the "V" chief of the Fifth Reich, indicating "V" for *vorshuchungs*, the German word for research.

As part of the introduction, Mike was supposed to hand Hartl the letter, along with his military discharge card. On the card was the name of the officer, Connelly, who had signed off on his discharge. This was to serve as his proof that he was Mike Muckerheide.

Mike left for West Virginia late afternoon on April 15, 1977, arriving in Morgantown in the early hours of the 16th after stopovers in Chicago and Pittsburgh. He booked into a motel for a few hours of restless sleep before rising early to meet Hartl at 5:00 AM.

Hartl was a large and ruggedly built individual, with a thick, hill country accent. There was a polished intellect about him, however, in stark contrast to his plain surroundings. It was obvious that he had not spent all of his time in the West Virginia hills. He had been referred to in Iwen's letter of introduction as "DRM" (der Reich marshal) — the title given to Hermann Goering in Nazi Germany.

After reading the letter, Hartl looked accusingly at Mike and asked, "How do I know that you're not someone else and that they

On the Laser's Edge...56

haven't killed the real Mike?"

It was an odd way to start a conversation.

"Well, I guess you'll never know that," Mike responded calmly, ignoring the wormy feel of apprehension crawling up his spine.

Hartl chuckled, then started walking away, towards the parking lot. "Wear different clothes next time," he said. "A suit and tie in these parts attracts attention."

After leaving the main highway in Hartl's Jeep, the two men wove their way into a remote, wooded hillside scattered with small, simple shacks that seemed exhausted of existence, weathered, paintless and sagging. As Mike caught glimpses of their thin, hollow-cheek residents, he thought he had never seen anyone so impoverished since wartime Korea.

Eventually, they stopped at one of the small houses, poised on a ridge, overlooking a valley. Mike followed Hartl up a set of groaning, wooden porch steps, through a creaking door, and into a small room that was dirty and dingy. Hartl's mother was there, but she barely spoke a word.

Hartl offered his visitor breakfast. Out of politeness, Mike accepted, trying to ignore the large roach-like insects that skittered across the table — "meals on wheels" he would joke in recollection. The bread was "old and stagnant."

Later, Mike was shown an underground room beneath the house filled with food and ammunition. It had been designed as a bomb shelter. Now, Hartl had plans for converting it into a laboratory for Mike. "Will it do?" he asked.

Mike knew the room would have suited the purpose, but he wanted to stall for time and said it was too damp. Hartl assured him that he would do whatever was necessary to create an appropriate laboratory, even if it meant building it from the ground up.

Mike had survived his trip unscathed. Two months later, he was off again, traveling to Colorado.

A second site being considered by Iwen was located near Fort Collins. His connection there was a man named Lt. Colonel Archibald Roberts:[20] an outspoken, ex-paratrooper who retired from the Army

---

[20] Lt. Colonel Archibald Roberts is the author of several books including *The Republic: Decline and Future Promise* (1975: The Betsy Ross Press); *Peace: By the Wonderful People Who Brought You Korea and Viet Nam*; and *Victory Denied.*

Code Word Tikal...57

with a bad attitude after 26 years of service. "He was very angry about America," Mike recalled. "Apparently, he had been a real warrior in Korea and he was angry because he felt that the U.S. had let him down."

His anger caused Roberts to put his heart and soul into the development of an organization called the Committee to Restore the Constitution (CRC). Iwen served on the Board of Governors for an affiliated group called the Wisconsin Legislative and Research Committee, Inc. Basically, they contended that a secret committee of bankers and anonymous insiders was controlling the U.S. government and was attempting to wrest local control away from ordinary citizens through property seizures, with the ultimate goal of replacing the U.S. constitution with a government that would leave citizens as mere serfs on their own lands.

In June, Mike and Iwen flew to Fort Collins to meet Roberts and to investigate a former training camp located in the Rocky Mountains. It had once been used by another organization of militants called the Minutemen.

On the night of their arrival, Mike and Iwen stayed at the Holiday Inn. Mike called Pat, as he did every night during his travels, this time using a pay phone in the lobby. He avoided his room phone since Iwen was paying for the trip and the bill would leave evidence of his calls. He arranged a secret code word in case something went wrong. He told Pat that if he used the word *chair* in a conversation, it meant something had gone wrong and she should contact the FBI as soon as possible. He never used the word, but devising the system had been wise because Iwen always followed him to the lobby, hovered nearby, and listened.

The next day while Iwen investigated the training camp, Mike was left behind to speak to another scientist. The man spent most of the time asking Mike technical questions — "sweeping my head," Mike said — about lasers.

The trip to Colorado wasn't so important for what happened there, as for the information it yielded during the flight home. Iwen, apparently now very comfortable with his conspiracy partner, confided to Mike the whereabouts of Posse leader Thomas Stockheimer. Stockheimer had disappeared in July of 1976 on the day he was scheduled to begin serving jail time for the assault of an Internal Revenue Service agent. When his abandoned vehicle was found on a

On the Laser's Edge...58

quiet country road, some of his supporters believed he had met with foul play at the hands of authorities. A federal judge, not receptive to conspiracy theories, issued a warrant for his arrest on bail-jumping charges.

On their flight from Colorado, Iwen showed Mike a recent letter from Stockheimer, addressed from Dulzura, California. In it Stockheimer said, "We're at war and have to remember we're at war." If Mike had any capacity for a photographic mind, he hoped for it now. He memorized what he could, then asked to see it again before they landed.

If there was an FBI follow-up in California, it never yielded an arrest. A couple of months later, however, Iwen told Mike that Stockheimer had been spirited away to West Virginia using an underground network of extremists and sympathizers. It was rumored that he was sometimes disguised as a woman. He was now living at the home of John Hartl, occasionally using an assumed name. When Mike informed the FBI, they asked him if he would be willing to return to Hartl's home in West Virginia to see if Stockheimer was there. He knew it was important and said that he would. Years later he would look back and realize, "My God, what a stupid thing."

CHAPTER EIGHT

*The prayers of both could not be answered;
that of neither has been answered fully.*
— Abraham Lincoln

Thomas Stockheimer was the first Posse member to thrust the Wisconsin organization into the spotlight. In 1974, he assaulted Internal Revenue Service agent Fred Chicken.

Agent Chicken had gone to the home of a local farmer in western Marathon County to review the farmer's tax records. A house full of men, including Stockheimer, met him there. Stockheimer threw a punch that struck Chicken; then, he shoved him across the room and into a chair. For the next hour, in a dimly lit room with a cache of rifles lined against a wall, the Posse leader lectured the agent about the illegality of the income tax and preached the doctrine of the Posse.[21]

Chicken was set free with a "God bless you" from Stockheimer who was later arrested and convicted for the assault, but only sentenced to sixty days in jail. Defiantly, he appealed and remained out of jail for the next two years, except for periodic arrests for other incidents in 1974 and 1975 on various charges of battery to a police officer, obstructing a police officer, and resisting arrest. As always, Stockheimer claimed his mission was to regain local control for the people, setting up a Posse style government to replace the government that he felt was illegitimately in charge.

Stockheimer had been on the run for more than a year. The FBI was anxious to renew acquaintances.

Mike left for Morgantown on Thursday, October 13, 1977. To Iwen and Hartl he used the excuse that he wanted to re-inspect the possible lab site he had seen there in the spring.

---

[21] The activism of Thomas Stockheimer and the incident involving Fred Chicken are documented in court records and news articles.

On the Laser's Edge...60

Mike had stopovers in Chicago and Pittsburgh. At the airport in Pittsburgh, he realized he was carrying a card containing sensitive information, including the phone numbers of FBI agents. He couldn't take the chance that Hartl or Stockheimer might frisk him and find the card. With a little time to kill between flights, he sat outside the airport terminal in an area among some trees, memorized the information, then burned it and proceeded back inside to board his commuter flight to Morgantown.

Mike was arriving earlier than he had told Hartl so he could meet first with an FBI agent scheduled to meet him at the airport. The agent didn't show and Mike was bothered and annoyed by the oversight. "Something is really wrong here," he thought to himself. It was a bad omen.

Someone else had watched him disembark, however, and was there to greet him. It was John Hartl. Mike realized then that the agent's absence, while bothersome, was fortuitous.

Mike and Hartl traveled again up through the hilly countryside. This time Mike tried to pay particular attention to the path they followed. This was West Virginia 1970s-style when pitted, unmarked roads led to many homesteads. There were no specific addresses, just rural route numbers. You had to be a local to find anyone, and many residents preferred the anonymity: it was a good way to keep out unwanted visitors.

Near the bottom of the hillside they crossed through a creek, then ascended into a heavily wooded area. Mike remembered the crude houses, the lush landscape, and the abundant supply of howling dogs along the way: nothing to make it stand out against the backdrop of any other hillside in West Virginia.

It was an overcast day and Mike couldn't tell north from south. He looked to the woods for help, and found none. The adage that moss only grows on the north side of a tree didn't apply here: moss grew everywhere, all over the trees, all over the ground.

During the ride, Mike mentioned that he knew about Stockheimer's travels to West Virginia. "I'd really like to meet the guy," he said. "Think it's possible?" Hartl was evasive at first, then conceded, "He's here, but I don't know about meeting him."

At the Hartl home, the two men sat in the small, dingy living room and talked; the West Virginian asking about Mike's recent research. Suddenly, a side door flew open. A short and stocky man bolted into the room. It was Stockheimer.

Code Word Tikal...61

Hartl made introductions and both he and Stockheimer talked about the coming revolution. Mike didn't contribute much, mainly asked questions and listened. He grew more fearful and anxious at some of the comments.

Then the conversation turned solemn as they led Mike outside. At the freshly dug hole they paused. "This is what we do to people who talk," one of them said, nodding towards the hole. Mike knew they were talking about him.

They continued on, headed towards a building countersunk into the earth and scrub brush. Hartl opened a trap door that creaked open to reveal a stairwell of railroad ties leading down to another door made with thick slabs of wood and hinged with heavy straps of iron. "Get down there" the shorter of the men demanded and Mike stumbled unwillingly down the stairs and into a cavernous room.

The door closed tightly, sending the men into total darkness. Mike thought of his wife, his daughter, his career and all the things he had dreamed of accomplishing. It was all going to end now. He reached into his coat pocket and felt for the familiar leather case that carried his well-used rosary of black onyx and silver and clutched it for comfort.

For an eternity, Mike stood in the blackened stairwell, his heart pounding wildly, heaving like an eruption. He didn't know how he could escape, and doubted that he'd ever get out alive, but he then decided : "if they plan to kill me, I won't go down without a fight."

Then it came, the sound of a hammer being drawn: *click*. It was followed in an instant by a bright light that burned into Mike's eyes.

Soon, the burning stopped.

Mike could still feel his heart beating. He was still standing. The *click* had only been the sound of a light switch. The room before him was bathed in a brilliant light.

"So, what do you think?" Hartl beamed enthusiastically. He was asking about the room. Stockheimer chuckled.

As with the light, it took a moment for Mike to adjust. He realized that the two men had been playing with his mind.

He was now relaxed, or at least relieved to be alive, and began to take in his surroundings, surprised by how much had been done to prepare the room for experimentation. It contained some lights and tables, and built into one wall, at about shoulder height, was a sliding window. A lab bench was positioned nearly even with the window. From there, one of the men suggested, they could test fire a laser into

On the Laser's Edge...62

the valley below.

"Looks pretty good," Mike had to admit.

Seeing the sophistication of the lab, however, was a disappointment to the Wisconsin scientist. The lab was ready. He wondered how much longer he'd have to continue his charade before the FBI would have enough evidence to arrest someone. Would he survive long enough to find out?

If there was ever a time that Mike felt totally alone it was now. Old doubts began to resurface. The FBI agent who had scheduled to meet him earlier in the day hadn't bothered to show up, so if something had happened to Mike, would the FBI have cared enough to come looking? And what about Iwen's friends with their science and military backgrounds? It was a bothersome connection and Mike wondered sometimes if the government, and the FBI, were somehow involved. He thought, "Maybe these guys are feeding information back to each other and now I can't trust any of them."

As he toured the underground facility, he began to plan for how he would end the madness himself. Assuming he would survive this visit, he decided, "If I did build a laser there, I'd built it with hydrogen, and I'd introduce oxygen into the stream of the laser, and by doing that, it would become a bomb and it would blow the place to Kingdom-come."

As Hartl prepared to drive Mike back to the airport that afternoon, Stockheimer took out a book of scripture and prayed. Then: "He threw his arms around me and gave me a French general's embrace... He gave me a bear hug and wished me well, that this would all come down, and it would be a grassroots thing and finally America would be saved," Mike said.

On his return trip to the airport, Mike again tried to memorize the trail, to find something out of the ordinary; and again, nothing seemed to connect. He flew back to Wisconsin feeling lost in the West Virginia geography, and frustrated that he might not be able to confirm Stockheimer's location.

In Wausau, Agent Page tried to coax whatever information he could from Mike's bewildered brain. It was not enough for Mike to tell him that the fugitive was staying at the Hartl home; the Bureau needed confirmation of where he had been and a good map of the property. The situation was potentially dangerous and the details had to be precise before attempting an arrest.

Code Word Tikal...63

They were making no progress until Page asked, "Were there any sounds? Any certain sounds?" It was that suggestion that triggered a memory. "There was," Mike recalled. "There was a sound like some kind of big, hydraulic, pile-driving machine. Every little while you could hear it crush or push."

Page recognized the value of this little bit of information. He immediately sent a teletype to FBI Headquarters and to the Pittsburgh, Pennsylvania regional field office which would oversee the Stockheimer capture. It included the caution that Stockheimer was armed and dangerous.

It was up to Pittsburgh to take the next step.

The operation to arrest Stockheimer was in the hands of FBI Supervisor George "Lefty" Stephan from the Pittsburgh office. Stephan was an old time criminal supervisor who had spent most of his career in Chicago before returning to his home area. Colleagues said he looked after his agents like a mother hen. Under his authority was Stephen Thompson, the Case Agent, out of the Fairmont, West Virginia office, which covered the Morgantown area. It was Thompson, a Texas native, who would lay the physical groundwork for the capture, feeding details back to Stephan for the overall operation.

Using as much information as they could from Page's description, the local agents in West Virginia discovered that the noise Mike heard was a car crusher from a scrap iron business in the area. It was enough to convince them that he really had been to the Hartl home. Now confident it was the fugitive's hiding place, Thompson decided to do a little early-November hunting in the Morgantown hill country. "I put on my hunting clothes and went up there with my trusty shotgun, and just acted like I was squirrel hunting," he remembered. He edged his way through the wooded growth within sight, but out of bounds, of the Hartl property. He didn't pay much attention to the squirrels that day, hunting instead for an outside entrance to the bunker. "We had concluded there were likely at least two [entries]: one inside the house and one outside," he said. Knowing where they existed would determine where agents would be stationed during the takedown operation.

Thompson couldn't clearly identify an entrance during his hunting expedition, so he enlisted the help of the State Police. Thompson and a Bureau photo clerk were flown over the area. Aerial photographs revealed a suspicious hump in the ground, which was thought to be a likely entrance to the bunker.

With the location well defined, a team was assembled for an official FBI visit to the Hartl home. It would include agents from the Fairmont, Charleston and Wheeling offices, and several members of the newly organized SWAT team out of Pittsburgh, along with uniformed officers of the State Police. The SWAT members were instructed to surround the house, approaching through the woods, while a couple of State Police cruisers and FBI agents would drive right up to the house. "We wanted a marked car and some uniformed officers, so if they did start shooting, it would be hard for them to say they didn't know we were law enforcement officers," Thompson explained. In all there were twenty to thirty law enforcement officers on the scene. "If you go loaded for bear, then you can kill a squirrel pretty easy," Thompson said. "That's kind of how the Bureau operated back then. It was better to have too much than too little."

The Bureau had its reasons for caution. For one, the Posse Comitatus didn't recognize any law enforcement beyond an elected county sheriff. That didn't put the FBI on the best footing. "We realized we weren't going to be on friendly ground," Agent Thompson said. They also knew about Stockheimer's assault of the IRS agent in Wisconsin, and had information that Hartl had worked at the Redstone Arsenal in Alabama as an explosives expert. In other words, one of the men had a history of blowing up, and the other one had the knowledge of how to blow things up.

The raid was set for the early morning hours of November 10, 1977. Tom Burgoyne, one of the men on the SWAT team, all in their 20s, remembered the day. He said he and his comrades crouched in the chilly, damp woods, watching the house until dawn, then moved in when they received the signal to advance. They burst into the rugged shack just in time to catch a man in his underwear disappearing beneath the floorboards in a coal bin. A pile of coal would have concealed the door if anyone had had the time to shovel it across. They didn't, and the SWAT team lifted the hatch and followed.

Mike had told the Bureau what they should expect to find underground: a large basement stocked with survival provisions. The reality was more than they expected.

"Above ground it looked like a regular three or four room West Virginia shack, but they had spent a fortune underground," Agent

Thompson described. "There were three separate rooms and they were concreted in, with concreted halls connecting them." A large stockpile of food and water made it obvious that someone was preparing for long-term survival in the underworld.

Tom Burgoyne described the bunker as a maze or a labyrinth. "There was a ladder going down," he said. "And we went into a square room. ... As I remember there were at least two tunnels that left this room. I went down one hall, like a corridor. It was a winding thing, like being down in a coal mine. I remember going down with some of my partners of the SWAT team, creeping down this maze."

Although there was electricity, no one knew where to find the light switches and used flashlights to guide them. "It was dark and there were spiders. It was a scary, bad thing," Agent Burgoyne admitted. "We were told to be very careful because... [Hartl] supposedly had been involved with explosives and there might be booby traps."

It was one of the most dramatic cases Burgoyne would ever work while on the SWAT team. "Some of the big cities had these guys, and college campuses were having a whole lot of problems in those days, but we didn't... [in] West Virginia. We were not used to these country boys being this kind of an offender," he said.

Not long after twisting their way blindly through the underground, SWAT members heard the shouts and sounds of a scuffle. "All I know is I heard a ruckus behind us and someone was yelling and we all turned around and backtracked, " said Agent Burgoyne. They found one of their agents from the Charleston office with his gun drawn and pointed at Stockheimer. He had stumbled upon the fugitive crouched in a corner. Stockheimer was unarmed and taken into custody without incident.

Upstairs, Hartl was arrested for harboring a fugitive. In a search of the property, the FBI found only a few ordinary weapons: shotguns and hunting rifles.

At the time of his arrest, Stockheimer was disheveled from his rude awakening and dressed only in his underwear. "Calvin Klein wouldn't hire him to model underwear, I can tell you that," Thompson said, adding that Stockheimer looked older than his 45 years. "Maybe he just had a lot of miles on him," he contended. "When you're on the run from the law, you're not at your best. Your entire existence is spent looking over your shoulder."

Stockheimer was allowed to dress, and Thompson and the Special

On the Laser's Edge...66

Agent in Charge took him before the magistrate, John Fisher, in Morgantown for his initial appearance on the fugitive warrant.

While feeling victorious, the agents' troubles with Stockheimer were not over. As they waited for the magistrate in his office, their prisoner, an epileptic, experienced a grand mal seizure. "We had handcuffed him and run the handcuffs through his leather belt," Agent Thompson recalled. When the seizure started, Stockheimer, "snapped that leather belt like it was a string, and fell to the ground." The seizure lasted thirty or forty seconds, and Thompson said: "It scared us to death. I thought, 'Oh, lord, if this guy dies, I'll be writing memos 'til I retire." By the time an ambulance arrived, Stockheimer had recovered.

He waived a removal hearing and was returned to Wisconsin in the custody of the U.S. Marshal. He was sentenced to an additional year in prison for bond jumping, in addition to the original six month sentence for assaulting the IRS agent, but spent only nine months in the federal prison at Sandstone, Minnesota: a mere hand-slap for all the bother. It would be only one of many more imprisonments in Stockheimer's criminal future, however. At least for Mike, there was finally one less extremist on the loose he'd have to contend with.

CHAPTER NINE

Courage is fear that has said its prayers.
— Dorothy Bernard

If Mike hadn't been so perturbed, he would have been scared. He was sitting in the passenger's seat of Iwen's dark blue Mercedes, sensing the agitation, and wondering "what's his problem now?"

Iwen's problem was that Stockheimer had sent a message from prison to "check out the scientist." His arrest and the timing of Mike's trip to West Virginia seemed a little too coincidental.

Mike had worried that the dots might be connected. A few weeks had passed without incident, however, and he'd grown comfortable. He'd been roused out of his comfort by Iwen's evening phone call and their rendezvous in the park. He'd rehearsed a way out, just in case, and now it was show time.

Iwen opened up his glove compartment and pulled out an automatic pistol. Fondling the handle, he pointed the barrel at Mike's chest. "Do you like that?" he asked.

Mike flinched, but remained calm. He hadn't spent the last year-and-a-half without learning something about Iwen and his friends: they liked to play games. Now it was his turn. "That's nice, Al. Think you can get me some iron like that?"

More than being scared, Mike was angry: angry that his life had been turned upside down; angry that he lived in constant fear for the safety of his family. He didn't know how far to trust Iwen with a gun, but he knew it wasn't the right time to do something stupid.

Iwen was a man growing paranoid, and not without reason. Though avoiding prosecution in his confrontation with the FDA, he had been ordered to cease production of laetrile. He was constantly worrying that federal authorities were keeping a close watch.

As well, someone had recently broken into the airport hanger and sabotaged his airplane, putting protein powder in the intake ducts of the engine and carburetor systems. He suspected the feds were trying to keep him grounded.

On the Laser's Edge...68

"You know, somebody's been talking," Iwen said. "The feds are breathing down my back. If they so much as think I might have helped Stockheimer, it would only give them one more reason to come after me."

"Somebody's been talking, all right. Someone in your organization is squeaking, and I'm not about to be a part of it," Mike snapped back. "If you want me to help with this plan of yours, then straighten things out." He had a hunch that Iwen wanted to believe him; without him, the laser plan would collapse. He seemed to be listening.

"Remember Al, there's three B's that will always compromise a plan: booze, broads and bucks," Mike continued. "Some of your friends have been fooling around with all three of them. Be careful who you trust."

Mike made sense, and Iwen recalled that Stockheimer had called into radio talk shows and provided his real name for mailing lists while out on the run. "He probably betrayed himself," he concluded. He lowered the gun and stashed it back inside the glove compartment. "You're a good guy," he told Mike.

For a man with such bad luck, this time Mike had been very lucky.

Iwen had boasted on several occasions that there was plenty of money available for his laser and laetrile plan. To prove it, on one occasion he showed off a large value of Krugerrands. Mike's orchestrated escalating of estimates for Iwen's expanding laser plans, however, was making Iwen sweat over how to finance it. He could no longer depend on laetrile sales in the U.S. He was reduced to seeking more creative financing, even if it meant stealing.

"Iwen wanted to see this ruby laser I had," Mike recalled, "so I took it out of my basement and took it out to a garage in Merrill where Iwen lived. He drooled over that. Then he said something about the big one at Stevens Point. He said, 'We're going to get that one even if we have to steal it.'"

The laser Iwen was interested in stealing was in the laser laboratory at the University of Wisconsin at Stevens Point. Mike had helped to design the lab and bring it to full operational status in 1974. The laser in question had been installed in 1977 and could have been modified for weaponry use. Measuring eight inches high, by eight inches wide, by three feet long, it could have been broken apart and easily stolen. Mike made sure that extra security was placed on the lab. In a sense of humor, some of the technicians installed a sign on the door:

"This room is guarded by two guys with a shotgun two nights a week: you pick the night."

Finally, in an all-out quest for financing, Iwen headed for Mexico and Central America in late August 1977. He was accompanied by two of his ultraconservative associates: his attorney and his business partner. Their itinerary included El Salvador, Costa Rica, Mexico, Honduras and Guatemala. During their journey, the secretive Iwen would often travel alone, never reporting where he had gone, or with whom he had spoken. He told Claussen that his travel companions knew nothing about the laser plan. He had left them with the impression that they were looking for a place to relocate his lucrative laetrile operation.

After returning to the U.S., Iwen was upbeat. He had accomplished one major goal on his trip: he had met a man in Guatemala who was interested in acquiring laser weapons for his government. He was Colonel Luis Federico Fuentes, the Deputy Minister of Defense for Guatemala. Mike could not have been more displeased with the news. If true, Iwen had found a way around the one obstacle he had thrown in his path: the lack of money. Any government, no matter how poor, always managed to find it when they wanted it. If Mike ever thought that he'd already ventured too far, he felt it now.

CHAPTER TEN

In the souls of the people the grapes of wrath are filling and growing heavy, growing heavy for the vintage.
  – John Steinbeck, The Grapes of Wrath

One day in Korea, Mike was walking alongside a Navy buddy as they crossed a field. They tripped over a spider wire hidden in the grass. Mike threw himself to the ground and the earth exploded with a shower of grass and stone. When he looked up, he saw that his comrade was lying still, in two pieces: his body in one place, his head in another.

Mike wondered, in a fog-filled moment, how to put the pieces of his buddy back together again, to make him whole. He was a medic. It was his job to help people. A moment later he realized he couldn't, and the helplessness haunted him the rest of his life.

He was sent back to the states to help deliver the news to the victim's family. It was just before the holidays, and just a few hours drive from Central Wisconsin, but the military would not give him a pass to go home to see his family. He was shipped back to Korea, back to the war, with no privilege to grieve. He would keep his grief as a closely guarded secret his entire life.

Some secrets are too large to keep; others too large to share. The secret of his grief was one that Mike would choose not to share; the laser conspiracy brought him one more secret that he dared not to share. "I talked to no one," he said about the conspiracy. "I couldn't tell who was who. I knew these [extremists] were everywhere. I'd hear people at work, people talking about the Posse, and I said nothing."

Carrying around such a heavy, mental load is it any wonder that he felt like Humpty Dumpty falling off his wall and grabbing at air?

His long hours in the laboratory were a convenient way to avoid confrontation, but too many things were happening now for Mike to keep that part of his life buried from Pat. She knew the sketchy details of his trips to West Virginia and Colorado, and now the FBI was relaying messages by phone to their home.

Pat Muckerheide was a very quiet, introspective woman, even around family, and rarely questioned Mike's outside activities. It wasn't because she was naive; it was because she was completely confident in his abilities. They had been married for more than 20 years and she knew the man she was married to: he was a good man, a faithful husband, a proud father. He was brilliant in ways that often amazed and amused her. He was also very opinionated and stubborn. She knew he would do whatever he felt he had to, regardless of her own opinions.

Even stronger than their marriage was their mutual faith in God and the Catholic Church. Neither wore religion on their sleeves; rather, it was something that was sown into their souls. They were Muckerheides, they were Catholics: it was as simple as that. They never missed a Sunday mass, Tuesday Novena or special church celebration. They offered prayers before and after every car trip, and before and after every meal as naturally as they breathed. Pat took great comfort in knowing that Mike sought divine guidance in his life, and so did she. If he had ever argued with God, he had buried it somewhere on the battlefield in Korea.

Still, it was a tense time in the Muckerheide household when Pat suspected some of the details she was denied. She knew what her husband was doing was potentially dangerous for his life, his family and his career. On one of his returns at 2:00 AM from a secret rendezvous, she asked him, "Well, who have you been out with tonight: the good guys or the bad guys?" It was a question that defined Mike's very existence at that point in his life: a man being torn in two by the forces of good and evil.

By the end of 1977, Mike was dropping to the bottom. He didn't know who he could trust. The continual meetings with Iwen, numbering in the dozens, were like a brainwashing. "Things happen to your head," he admitted. "It's really confusing. You're trying to live right and they're trying to bend you the other way." The episodes in West Virginia and Colorado, he admitted, "were turning muscle into Jell-O."

He wanted nothing more than to pursue his scientific career; at the same time, he was too entangled in the laser conspiracy to bail out.

His doubts about the government lingered. It had been a year and a half since meeting Iwen, and the FBI had done nothing to take him out of circulation. "I didn't know what to do anymore," he said. "The Bureau was lagging. It was not moving. I thought they should have been moving against these guys. I felt that acutely." It was one more reason to push him towards the thought that the government was either involved with Iwen, or involved with some kind of cover-up.

The next logical leap was the notion that hooking up with Iwen might not be so bad. Even if he didn't particularly like the man or his politics, he recognized his intelligence. Their conversations were at times intense and thought provoking. He liked that. And, Iwen did not question his expertise in lasers as some people did. While Mike was enjoying better relations at A. Ward Ford Institute, the politics of research never stopped being difficult and emotionally exhaustive.

The subject of laetrile also interested Mike. Pat's cancer had finally gone into remission (agents would wonder, only half-jokingly, if she had been cured with the zap of a laser) and the experience was driving Mike's interest in finding an effective treatment for the disease.[31] Never mind that he would soon realize that science didn't support the claims for laetrile, in the beginning it was worth considering.

Also nagging Mike was that part of him that never wanted to go back to the depression years of his youth. A part of him really did want the money.

In his conversations with the chemist, he knew the laser plan was creeping towards reality. There was lots of money to be made if they could connect with the right foreign country. Better yet, that country could provide them safe haven from prosecution. He was not against making money and providing his family the standard of living that being a scientist in a small city could not provide.

Mike always believed that if he prayed for an answer it would be given to him, so he prayed often and fervently for guidance. Impatiently he waited for an answer that did not seem to come. Eventually he would think: *Maybe I'm looking at this all wrong.* In his own mind, alone with his thoughts, there were days, more and more of them, when he made the turn, and committed to Iwen. *Maybe,* he thought, *for reasons I don't understand yet, that is what I'm supposed to do.*

Then, finally he decided: he would do it. He would work with

---

[31] "Wausau researchers study lasers on cancer", Geri Nikolai, *Wausau Daily Herald*, October 29, 1973.

Iwen. He would build the weapons Iwen wanted.

Mike eased into a sense of calm after his decision. It was so easy now, he thought. He'd been swimming upstream against the current. Certainly, it would be much simpler to get rid of the FBI than to get rid of Iwen. He could walk away from the Bureau and they probably wouldn't even care.

After his decision he paid a visit to the Wausau FBI office. As he approached the building and climbed the stairs to the second floor, he didn't know how he was going to start pulling away, how to dislodge the Bureau from his life. It would have to be subtle. Once he was successful, he'd go to Iwen and give him good information, usable information, for developing laser weapons. They would work together as a team.

Page was working at his desk when Mike arrived that day. He looked up briefly when the scientist entered, grunted his hello and scribbled something more before closing the folder and shoving it aside.

"Take a seat," he offered Mike.

"No, that's all right. I won't take long," he said pointedly. He didn't want to waste any time with small talk. "I just need to know something. Why aren't you guys telling me anything? Why can't you tell me what the hell's going on with this case?"

Page didn't notice anything particularly unusual about Mike's demeanor that day. He knew he could be a straight-shooter at times, firing questions and expecting immediate answers. He did sense that he was in no mood for a runaround.

"Mike, we can't tell you a lot of things" he started just as pointedly, looking squarely at Mike, eye to eye. "If we told you, and by mistake you said something to the wrong person, it could lead you into some very big problems. We don't know what these guys are capable of. They're playing a very dangerous game."

Mike hesitated. His tense body relaxed. Maybe it was the way the agent said it, or just Mike's willingness to hear it, but in that moment, something made sense. He finally understood why the FBI was so unwilling to tell him anything, and it made a difference. Mike had been using the same excuse for not telling his own family about many of the things that were going on in his life: he kept quiet to protect them.

If he had been waiting for a great epiphany, a moment when the angels sang, this was it, from the lips of the taciturn Page.

As Mike took a seat, the two men continued to talk and Mike's anger began to melt.

In the end, Agent Page left him one more valuable piece of information: "Scientists usually don't tip," he said, referring to the devotion of scientists towards their fields of science and their tendencies not to be tempted into doing something that might discredit or compromise their science. It instilled in Mike a renewed pride, in and out of the laboratory. From that time on, even when faced with the frustrations of research politics, the condescending attitudes of colleagues, and his inability to finance so many of his ideas, he firmly resolved never to discredit the laser community by tipping.

It was a revelation that the agent who rarely told Mike anything had just told him everything he needed, and Mike left feeling a great burden had been lifted. He would return to live his double life for the Bureau.

The end result was expected, though, for at his core, Mike was not a man motivated by money. The Bureau had offered more than once to reimburse him for some of the expenses of being an informant. He always refused. "I remember broaching the subject one time and did he want to get reimbursed, and he was mortally offended," Agent Southworth recalled, "I said, 'Mike, you're putting a lot of time into this and driving your car around. We're more than willing to pay you.' And he said, 'No, no. Public service.' He was just kind of offended that I asked."

No, it wasn't money that motivated him. It was lasers. Laser science was so much a part of Mike's life then that he lived it and breathed it. It was such a great source of fascination and inspiration that he could lose himself in the science and regain some sense of sanity. In 1977, despite his ominous encounters in West Virginia and Colorado, and meetings with Iwen that took place almost weekly, Mike threw himself into his work.

In addition to updating the University of Wisconsin - Stevens Point laser lab, he was working hard on projects at both Detroit and Wausau, primarily in the areas of heart disease and arthritis. For the Electro-Optics/Laser 77 Conference & Exposition that year, he teamed up with Dr. John Goldman (son of Dr. Leon Goldman), and a Dr. E.C. Muehlenbeck to present a paper on "Laser Treatment of Rheumatoid Arthritis." At the same conference, he teamed up with

Dr. Drake for a presentation on "Use of Lasers in Revascularization of the Heart".

The patent Mike had been working on with Dr. Uecker at the time of Uecker's death and with Battelle Laboratories came to a conclusion in 1977. The patent papers were filed that fall.

And, in 1977, Mike received the *Outstanding Citizen Award* from North Central Technical Institute at Wausau for his volunteer assistance in developing a two-year laser technician program at the school. At long last, he was beginning to feel some success at blowing off his critics. He was being accepted by the laser community at large and was finally moving forward in his reputation and his research.

While his successes with lasers, and Agent Jack Page's occasional advice, helped to calm Mike's growing frustrations, he was shaken by the fact that he had almost been persuaded to join Iwen's efforts. No, doing right wasn't always easy, but it was always necessary and he knew that Iwen's way was not right. It had been a test of himself that he had almost lost, and it scared him.

CHAPTER ELEVEN

Lasers are born just lasers; they are made good or bad by the people who use them.
— Mike Muckerheide

Mike was cleaning his garage one day when he found a large, glass lense. "You want to see something interesting that can be done with glass?" he asked his 10-year-old nephew, Dave Barwick, who was visiting that day. In the driveway, he held the piece of glass about a foot off the ground, angled to the sun. A small, bright dot of light appeared on the driveway below. As he raised the lens the dot grew, and when he lowered the lens, it shrank. "How does it do that?" young Dave asked in awe. Mike explained how the light waves from the sun were going through the glass, and because it was curved, it would bend the light.

Then Mike placed Dave's hand carefully under the lens so he could feel the heat of the light, letting him know that it would burn if he got too close. "He coached me on safety and responsibility and how light could be used for good things and bad. Then he gave me the lens on a promise that I would use it carefully," Dave recalled.

The nephew enjoyed playing with his new "toy." When he saw an ant crawling across the pavement, he adjusted his aim, placed the bright dot on the ant, and in a second, the ant exploded. "All that remained was what looked like a tiny piece of charcoal, and the smell of the burnt ant rising through the air," Dave recalled. After the discovery, Dave roamed the driveway burning leaves, small twigs and more ants.

Years later, Mike and Dave would talk about that simple experiment and ponder: If a small crude light could totally destroy an ant, vaporizing it to black ash, what will happen one day when technology advances, as it surely will, to replace the ant with an airplane, and the airplane with a building, and the building with a man? In future warfare, will there be nothing left in a flash of light, but a tiny piece of charcoal and the smell of something burnt?

During the latest U.S.-led wars in Afghanistan and Iraq, most people likely thought they were seeing the latest and greatest in battle technology. Indeed, there were laser sights, laser range finders, satellite communication lasers...on and on. It was all very impressive, and the precision of the weapons gave viewers comfort. There was less collateral damage and the bad guys could be picked off, while the surrounding crowd of the good guys were left standing.

Mike said, however, that what people were watching was old technology getting flushed out of the system. "The military is getting rid of their old inventory," he said. "Getting rid of it to make way for the new. There's some bad stuff out there. Some really bad stuff coming."

That's the problem with scientists: they often shatter illusions.

Most scientists don't concern themselves much with the politics of their work and it can make them an easy target for foreign nations, rogue groups or lone wolves who thirst for their knowledge. Mike was always conscious of the politics: always aware that, either for good or bad, somebody always wanted what a scientist had.

By the spring of 1978, Mike had been caught in the tug of war with Iwen for two long years. Iwen's want list had shifted; once wanting a sophisticated weapon mounted in an airplane to start a war, he had modified it to a more feasible laser handgun he could sell in volume to a foreign country. His enthusiasm over finding a Guatemalan official who wanted what he had to sell, was giving his plan legs. Mike needed someone in the FBI to understand the urgency before it moved too far ahead.

Agent Page had instructed Mike to meet him at the grandstand at the Marathon County Fairgrounds on the west side of Wausau, inside Marathon Park. It was very familiar territory to Mike who, in his youth, had spent a lot of time there sailing his homemade battle ships in the park pool.

Mike pulled up to the log parking buttresses, got out and walked across the grassy field to the large grandstand. Up in the bleachers Agent Page and a second man were waiting.

Mike knew that Agent Page was leaving Wausau for a new assignment in California. This day, Mike would be meeting his replacement, Tom Burg, and he was nervous about the change. Despite a bumpy beginning, Page had been with him through some difficult times and the case was heating up. He had been fortunate enough to have two good agents to confide in; would he be so lucky a third time?

Agent Burg was a man in his mid-30s, six feet tall, with hair mowed short to dark stubble. He still had that scrappy look of his high school football days, with a fit and blocky frame. In his senior year at Cincinnati, Ohio, playing tackle, he was the largest member of the football team at a modest 200 pounds. What the team lacked in brawn, they made up for in brains by "outsmarting our opponents," he contended. The Greenhills Pioneers ended their season with nine wins and one loss, placing them 15th in the state. Remarkably, both tackles and the place kicker went on to have careers with the FBI.

A supervisor used to say of Burg: "He never met a case he didn't like." He found the long, hard hours an interesting challenge, yet also managed a successful personal life. Wisely, he had married a former FBI secretary who understood and accepted the occupational hazards of his career. They were making plans for a family.

A pleasant, laid-back personality concealed a history of dealing with bank robbers, arsonists, rapists, and murderers: the usual fare for Bureau agents. His lack of flamboyance would cause his daughter to one day satirize him as "Mr. Excitement."

Burg was a ten-year veteran when he moved to Wausau. He had started out in Laurel, Mississippi; headquarters of the violent White Knights of the Ku Klux Klan of Mississippi at a disruptive time in the group's history. He got his fair share of lessons in southern social history and politics. His biggest case there was helping during the trial of those charged with the death of black storekeeper Vernon Dahmer in Hattiesburg.[32]

He was later transferred to Chicago, where he was assigned to the Fugitive Squad and where he met his future wife, Pat. Then it was on to Waukegan, Illinois.

While in Waukegan he was given the opportunity to sign up for a permanent position anywhere in the United States. Selection to one's Office of Preference was based on seniority and the availability of an open position. Agent Burg's own preference had been somewhere out

---

[32] Sam Bowers, a former Ku Klux Klan Imperial Wizard, was tried four times for ordering the murder of Vernon Dahmer, each one ending in a mistrial. On August 21, 1998 he was tried a fifth time and convicted. He was sentenced to life in prison.

west, but Wisconsin was his wife's home state, and also a favorite. Wisconsin won.

It wasn't difficult to find an opening in Wisconsin since it wasn't a state that attracted many applicants. "Agents don't like lousy weather and high taxes, and Wisconsin had both," Burg noted. He was given his permanent assignment as a resident agent at a two-man office in Wausau. He, and a partner who changed every few years, were responsible for the northcentral part of the state. It was a decision he never regretted. "For a change, they put a round peg in a round hole," he contended over twenty years later. "I was in the perfect place for me, and I think in the best place suited to my abilities for the Bureau."

Agent Burg had originally planned to be a chemist, receiving a bachelor's degree in chemistry at Wilmington College in Ohio and a master's degree in chemistry at the University of Idaho. He hired on with the FBI thinking he would spend his life holed up in a Bureau laboratory. As part of his required work experience, however, he discovered he liked fieldwork. "I enjoyed it," he said. "I didn't realize I was such a people person, but I discovered that I was."

He was also the kind of person who liked the diversity of a small office. "A rural agent has to be a real country doctor," he noted. "You can't specialize."

The diversity worked against him in the beginning. There were a lot of cases pending and he needed to become familiar with all the notable players of all the notable cases before Page left. As part of the routine, he would drive around to see where people lived and worked and what their daily habits were.

The day he met Mike, he was too busy getting acquainted with all aspects of his job to really notice much in particular. He did know that the laser case was entangled with other cases involving Posse members and laetrile, and that Mike was somewhere stuck in the middle of it.

At first glance, Mike liked the new agent, but he was skeptical. "I had gotten used to Page," he said. "Page was very laid back, and all of a sudden this guy, Tom, he was real aggressive. He asked a lot of questions."

Neither man recalled much more about their first meeting except a question Page raised. It was about the money Iwen was attempting to raise for the laser scheme. "Page asked me if I thought Iwen might try robbery to finance the plan," Mike said. "And I can remember Tom said, 'You mean, this guy would try to rob a bank?' and that kind of startled me because I hadn't even thought about that." Instead of

On the Laser's Edge...80

easing Mike's mind, the agents had just given him one more thing to worry about. It was an awkward start for the new agent and the weary informant.

Every couple of weeks, Mike and Agent Burg would meet and they would discuss the case and get to know one another. They quickly found common ground. By coincidence, they both had married executive secretaries nicknamed Pat. They also shared a scientific background. To Mike's surprise, Burg understood the scientific mind and understood the bizarre possibilities of laser science unlike any agent Mike would ever meet. "I think most FBI people are accountant types," he said, alluding to a move by the FBI in the 1970s to recruit accountants when the Bureau decided to focus more of its attention on white-collar crime. "Tom had chemistry," Mike added. "He's scientific minded. He could understand. Maybe for a bean counter it was quite a jump to understand the laser case, but not for a scientific person."

"I had been around brainy types before," Agent Burg explained. "Because of my own scientific background, we just hit it off."

By the time Burg began his new job in Wausau, the laser case was starting to boil. The reason was the introduction of a new figure in the case.

Iwen had started mentioning the name of a respected politician from West Bend named James R. Lewis. People liked him and thought he was going places. Iwen and his business partner, David P., had spent time in his office waiting to testify at laetrile hearings in March 1977.

Lewis was almost 42, meticulous and polished, with a perfect, piano-ivory smile. He had graduated from Rufus King High School in Milwaukee, attended Moody Bible Institute in Chicago, and worked as a sales representative for the Wisconsin Bridge and Iron Company at the time he first ran for state office in 1972. He was also a part-time deputy sheriff.

He was married with two children when he first entered state politics, then divorced. With a salesman's history, he had a salesman's personality: pleasant and outgoing. It was combined with an ample sexual appetite. In Madison, he quickly gained a playboy's reputation. One of his political colleagues said that he "screwed anything that wiggled;"[33] however, he had worked into a steady relationship with

---
33 Confidential source.

one of his young and attractive campaign workers, Deborah Batzler.

Lewis and Iwen had a few things in common, most notably their ultraconservative values. Though a Republican, Lewis leaned so heavily right that he sometimes slipped off the party platform. He had found it advantageous to keep his more extreme views to himself and was a respected member of the Assembly who was being eyed for a top-level position in the state.

Iwen kept dropping Lewis' name into conversations throughout 1977, and the beginning of 1978. It soon became apparent that he had found a new friend and was adding him to his exclusive circle of confidants. If there was ever any doubt of that, it didn't last long after Lewis accompanied Iwen on a two-week trip to Central and South America.

Lewis had acquired his passport to travel out of the country in February '78. The trip was sidelined with Iwen's ongoing troubles at his Manitowoc lab. That month it was again raided by FDA officials and laetrile supplies were confiscated. Iwen appeared in court to face contempt of court charges, refusing to comply with an order to reveal the names of people he had sold laetrile to, and to reveal the hidden location of laetrile production supplies. Finally, in April, still a free man, he set his sights south, to the Tropic of Cancer.

Iwen and Lewis left from O'Hare Airport in Chicago on April 10 and arrived in Guatemala City that same day. Their first stop was the U.S. Embassy.

Over the course of the next two weeks, Iwen and Lewis traveled to Nicaragua, Paraguay, Bolivia, then back to Guatemala. They met with several officials during their travels, including Colonel Fuentes, a contact Iwen had made on his previous trip to Guatemala. Representative Lewis further paved their way by contacting the office of U.S. Senator William Proxmire. Had Proxmire's office staff known the underlying motives, they likely would not have been so accommodating; acting out of simple courtesy to a state legislator, they had sent notice to various embassies to assist Lewis in meeting officials.

On Iwen's return to the states, both Mike and Claussen noticed that their well-traveled comrade was in a good mood.[34] Things had gone well in Guatemala. Colonel Fuentes, who had been on-again, off-again to Iwen's proposed business dealings, was definitely on-

---

[34] Sworn statement of John Claussen.

again with the appearance of Lewis. The state representative lent credibility to the plan, and promised to open doors in a way that only someone with real political connections could.

Poor, beautiful Guatemala. It is a destitute country rich in assets. The problem remains that most of its assets are owned by a wealthy few. Approximately 75% of its population lives in poverty, controlled by a government that is either generous or brutal depending on its leader of the day. Its land is fertile, its temperatures tropical, making it ideal for growing coffee, sugar, bananas and illegal drugs.

The travel industry says Guatemala is a great place to travel cheaply, if you travel carefully; it might cost you your life, or at least your luggage, if you don't.

One of the must-sees is the ruins of Tikal. The ancient, religious city is set in a lush, remote area of the Northern Plain and is part of what is left of one of the greatest civilizations of the Americas, the Mayans. Colonel Fuentes had flown a *National Geographic* team to the spot in the early 1970s and the product of their visit was featured in the magazine's December 1975 issue.

Civil war erupted between the government and civilian guerrilla groups in 1960. The United States was not entirely blameless for helping it along, supporting the overthrow of a president who was implementing social reform by taking from the rich and giving to the poor.[35] Subsequent leaders received mixed reviews, with brutality synonymous with some of them. General Kjell Eugenio Laugerud García was president at the time Iwen and Lewis were trying to court the Guatemalan government. His election was controversial and his support of a civilian death squad condemned by Amnesty International.

It wasn't only the government that supported terror of its own people: some of the left-wing guerrilla groups were no better. After the war ended in 1996, however, a Commission for Historical Clarification reported that the Guatemalan army was responsible for 93% of all arbitrary executions and forced disappearances in the

---

[35] Reasons for U.S. intervention vary depending on the source. The U.S. government position is that it feared Communist influences in the actions of then-president Col. Jacobo Arbenz Guzman. Cited by several other sources is the position that the U.S. was primarily interested in protecting the rights of the United Fruit Company, which was a major landowner.

country during the thirty-six year war; only 3% were committed by guerrilla armies. Of the 200,000 citizens who died, 83% were Mayan.

Except for the civil war, and in large part because of it, Guatemala seemed like a land of many opportunities for certain enterprising Americans.

CHAPTER TWELVE

*Our scientific power has outrun our spiritual power.
We have misguided missiles and misguided men.*
— Martin Luther King

If scientists didn't have the right to free speech, most of them would end up in jail. Their ideas are often considered bizarre and even blasphemous to outsiders. Much greater minds — scientists like Copernicus and Galileo — all had their problems with the populous, but were eventually able to overcome their critics. For the average scientist, however, most of their ideas never go anywhere, so it explains in part why FBI management was reluctant to rush to take Iwen and Mike too seriously. With the addition of a politician and a foreign military leader in the mix, however, a few more heads in the Bureau started to turn. No longer were the conversations between Iwen and Mike just two indefatigable brains pondering futuristic possibilities, now there were two people with power and influence buying into the idea: Representative Lewis and Colonel Fuentes. *That* they could understand. Confirming the connection and seeing how serious they were was the next logical step.

Unfortunately for the investigation, Iwen remained elusive about Lewis. One moment he was preening over his success at lobbying the legislator; the next moment clamping shut the flow of information.

Realizing that getting Lewis to show his face would help move the straggling case forward, Agent Burg had to come up with a plan. He approached Mike with his dilemma. "Maybe Lewis is involved with this," he explained to Mike, "but we can't take Iwen's word for it. We need to find a way to flush him out into the open."

To Mike's relief, Burg was the type of agent willing to explain things. The two had been talking on a regular basis for the past month, and were already showing promise of working well together. If Burg wanted Mike to do something, it came with a reason. Mike clearly understood the reasons for wanting to get Lewis on the scene.

Iwen, however, wasn't cooperative. He ignored Mike's prodding.

Exasperated, Mike told him, "Stop playing games. I can't afford to waste my career on a phantom partner."

Eventually, Iwen conceded.

It was a pleasant day in early June; a nice day for a picnic in the park. Iwen flew his plane to Madison, picked up Representative Lewis, and returned to the municipal airport in Wausau. A short distance away was Radtke Park. Around a picnic table, eating hamburgers from a nearby restaurant, Mike and Claussen were introduced to their new associate.

He seems a little uptight, Mike thought as they shook hands. He relaxed as they talked about his flight, the weather and their careers.

If there was going to be time when Lewis couldn't handle the conversation it would certainly come up when they turned to religion and politics.

"I'd like to blow the 1313 building in Chicago off its foundation"[36] Iwen said.

Lacking the hesitations of a newcomer, Lewis shot back: "What I'd like to do is take a plane and fly by the control tower of O'Hare Airport and beam a laser on the tower and blow the whole thing up." Then he said he would leave the country knowing that a nuclear war was on its way.[37]

If attacking Chicago hadn't fazed him, maybe Iwen's plan for attacking Cuba would. But after Iwen told the group that he wanted to install a large carbon dioxide laser in an aircraft, attack Cuba, then lay the blame on the CIA, "So that there would be a Soviet retaliatory strike against the U.S."[38] Mike didn't notice any change in Lewis' expression; no look of shock. Instead, the chemist and the politician started talking shop, contemplating where to set up business in Latin America. Lewis said he was expecting a call from Colonel Fuentes in a couple of days. Even if they couldn't arrange a business of their own in Guatemala, certainly they could sell lasers to the country for night fighting.

The hour long meeting was one of the most significant hours of the case. No longer did the FBI have only a wishful-thinking chemist with oddball ideas, and a laser scientist who was cooperating with

---

[36] Sworn statement of John Claussen.
[37] ibid.
[38] Muckerheide testimony to the Grand Jury, May 2, 1979.

On the Laser's Edge...86

authorities, they now had a public official looking to do some private and illegal business with a foreign country. It was something they could work with.

Lewis' arrival provided the FBI with the break they were looking for. Maybe laser science fell through the legal cracks, but not corruption in public office. This was something the Bureau was familiar with and had even been encouraged to focus its energies on.

Agent Burg was responsible for taking the next step. When he felt confident there was enough evidence of criminal activity to pursue charges, he turned to his supervisors in Milwaukee. He compiled a memo that asked for the case to be opened as an active investigation, moving the information out of the Informant File. Now the Bureau could do more than just gather third-party reports.

To Mike's way of thinking, a lot of laws had been broken and justice was overdue. The FBI didn't yet see it that way. Ultimately, one of four things happens to an FBI file: the information is channeled into an existing case file to help with an ongoing investigation; it becomes substantial enough to open a case file of its own; or it goes into a *dead file* or a *zero file*. Dead files contain information that isn't useful to building or supporting a case. Zero files contain information important enough to keep active, but with no obvious connection to any existing investigation.

Mike's information had started out in the Posse file, along with information about other extremist activity. There didn't seem to be any other logical place to put it since new rules at the Bureau — post-J. Edgar Hoover — had agents cautious about jeopardizing free speech or expression. The fact that Iwen's actions were partially motivated by political and religious beliefs, though extreme, was seen as a complication, along with the un-legislated science of lasers. The Bureau had obtained exemptions from the rules to allow them to continue investigating ongoing cases in the Christian Identity Movement and Posse Comitatus, but were reluctant to open a separate file for the laser case until enough evidence for filing charges came through the backdoor by way of informants.

There were five squads in the Milwaukee office: One Squad was the Special Agent in Charge; Two Squad was the Assistant Special Agent in Charge, also responsible for organized crime investigations, and beneath that were three more squads, each focusing on different categories of crime and headed by Field Supervisors. Three Squad was focused on white-collar crime and Four Squad on reactive crimes like bank robberies, interstate trafficking and blue-collar crime. It was Five Squad, supervised by Ray Byrne, that handled a hodgepodge of cases generally centered on issues of domestic security, civil rights, extremist groups, and foreign counterintelligence. It was on his desk that the case landed. From this point on, it would be his responsibility to maintain some day-to-day knowledge of how things were progressing.

After discussing it by phone, Agent Burg traveled to the Milwaukee office to officially open the case file. He met there with Supervisor Byrne and Special Agent in Charge Jerry Hogan.

Agent Burg had gotten along well with the affable Byrne in other cases, but to his own way of thinking, the older agent was too preoccupied with crossing T's and dotting I's. He had a frugal reputation as well, with a monotonous wardrobe of gray flannel suits to match, and hated splitting a restaurant tab. It was his indecisiveness, however, that caused Burg problems. In one instance, Burg became very frustrated by the supervisor's unwillingness to render a decision. He ranted and raved to his partner Richard "Zeke" Szekely, "How does this guy know what color of shirt to put on in the morning?" to which the dry witted Szekely replied, "He only owns white shirts."

Part of the problem between the two men was simply their age difference. Byrne was an old-school agent of about 60 who was finishing out his career. Burg was still young and enthusiastic and just hitting his stride. Byrne didn't know what to think about this new kid on the block with all his talk about lasers. Maybe the young agent wasn't getting the story straight, or he was just too gullible.

If Supervisor Byrne gave Agent Burg pause, then SAC Hogan gave him promise. Hogan was a gruff, cigar chomping, Irish New Yorker in the prime of his career. He had gotten acquainted with Agent Burg earlier when they shared a death threat from a bank robber. Burg was still working in Illinois when the case crossed state lines by way of the robber's girlfriend. Burg got the girlfriend to talk, leading authorities to a large stash of stolen and marked $5 bills. Before it was over, the robber threatened to kill Agent Burg, SAC Hogan and the Assistant U.S. Attorney in the case, but he eventually gave himself up and went to jail. The incident provided an unexpected dividend: Hogan was helpful in getting Burg his requested assignment to the Wausau office over a year later.

On the Laser's Edge...88

When Agent Burg delivered his memo on the Lewis case to Milwaukee, SAC Hogan reveled in the fact that the Wisconsin Bureau could finally lay claim to a public corruption investigation. Wisconsin had a reputation for being squeaky clean and skeptical supervisors in D.C. were always pressuring Hogan about the lack of corruption cases. With the appearance of Lewis, the joke at Headquarters was, 'Hey, some fool out there in Wisconsin has finally found some corruption!"

After Hogan, Byrne and Burg discussed the situation, they decided that the efforts of Iwen and Lewis to deal in arms with Latin America was a possible violation of Neutrality Statutes, which prohibited a private citizen from carrying on policy with a foreign government. Byrne was still skeptical about the lasers, but willing to take Burg's memo and give it his required signature to open a new file in Milwaukee. It was then passed on to the clerk where the case was assigned a number and the file officially created. From then on, everything about the file would be kept physically uniform with all other files in the Bureau, down to its two-holed, top-bound pages. Additions would be sent internally via a holed envelope nicknamed the *Swiss cheese envelope*.

The identification number assigned to the case was an incredibly low number for 70 years of FBI history of Neutrality Matters in Wisconsin. The Bureau's national file number was also extremely low.

Once the case was opened, Agent Burg began discussions by phone with the U.S. Attorney about possible charges. When charges looked promising, he headed to the state capital to meet with him personally.

CHAPTER THIRTEEN

*Time will bring to light whatever is hidden; it will cover up
and conceal what is now shining in splendor.*
— Horace

As a newcomer to Wisconsin, Agent Burg had not yet met the hard-driven U.S. Attorney Frank Tuerkheimer. He had heard plenty, however, about the prosecutor's reputation from other agents. It was not very flattering. It was said that he had a low opinion of the FBI and could be impatient and uncooperative. A lot of agents didn't like him, and Agent Burg often heard him referred to as, "that goddamn Tuerkheimer.'" Burg decided to reserve judgment.

Bringing the U.S. Attorney on board early was not standard procedure for all agents. Many agents would throw a report on the prosecutor's desk and walk away, leaving it up to him to sort out all the facts. Agent Burg, on the other hand, had a theory about prosecutors: he believed that keeping them up to speed personally helped save time; kept the process moving smoothly; and improved chances for winning cases. Sometimes, it even made them more benevolent towards agents.

The prosecutor's small office was located on the second floor of the Madison Federal Building, a block away from the capitol. It was a Depression-era structure built from stock plans that the government utilized over and over again throughout the country. Roanoke, Virginia had one just like it. So similar were they that a visitor could walk in off the streets of Madison and feel they had arrived at the facility for the Western District of Virginia.

What the agent noticed first was the amount of limited parking for the large complex. Having already passed the nearest municipal lot some distance away, Burg did what he was accustomed to in other cities: he found a convenient spot in a no parking zone along one side of the building, put his official, red bubble light on the dashboard, and strode into the building through a side entrance.

On the Laser's Edge...90

Burg knew he'd be spending a lot of time in this building in the coming years of his career, so he made careful note of its layout. The Madison Post Office occupied the main floor. The federal offices were located upstairs.

Burg climbed up two, narrow flights of white marble steps. At the top, he was confronted with a U-shaped configuration of hallways and rooms. The long hallway to his right led to the U.S. Marshall's office and a prisoner cell for short-term holding, the office for the Clerk of the District Court, the Western District Federal Courtroom, and to another hallway leading to the Grand Jury room.

He chose instead a hallway to his left. Walking past the two small offices for supporting assistant attorneys, he headed straight into the reception area for the Western District U.S. Attorney.

There were two receptionist/secretaries sharing the large room: one for the First Assistant U.S. Attorney, whose office was located directly ahead, and one for U.S. Attorney Frank Tuerkheimer whose office was located off to the left.

There was nothing glamorous about the reception area. It included a couple of desks, a few extra chairs, and a rotor of case files. It was designed in classic government-issue woodworking and decor. It had large, tall windows and globe lighting fixtures that hung from the ceiling on long, metal poles.

Even to Agent Burg, who wasn't accustomed to much office glamour, it looked cramped and uncomfortable. The single item of class, he noted, was an antique dictionary stand.

Agent Burg introduced himself to Tuerkheimer's secretary.

Ruth Larson replied, "Nice to meet you, Mr. Burg. Welcome to Madison. I'll let Mr. Tuerkheimer know that you're here."

Burg was a few minutes early for his appointment, but Tuerkheimer soon appeared out of a side room to escort him back into his office. He was a tall and wiry man, with a thick shock of dark hair. He was confident, decisive, with a no-nonsense personality.

At 39, the native New Yorker had impressive credentials. He had received his law degree from the New York University Law School in 1963, graduating cum laude. He started his career as a law clerk to Judge Edward Weinfeld in the Southern District of New York (1963-64), taking a detour to Africa in 1964-65 to serve as Legal Assistant to the Attorney General of Swaziland. Upon his return to the states, he became an Assistant U.S. Attorney for the Southern District of

New York, and continued to gain notice among his peers and superiors. During the Watergate years, he was appointed Associate Special Prosecutor in the federal investigation and worked as the lead attorney in the case of *United States versus Connally*.[39] In 1977, he was appointed to his post in Wisconsin's Western District by President Carter as a reward for his Watergate work.

Tuerkheimer's office was comfortably large, with a long, wooden conference table extended out in front of his desk, forming a T. He motioned for Agent Burg to take a seat at the table, and joined him.

There was little time for small talk in either of the men's hectic schedules. Burg plunged into explaining details of the case: the years of Mike's involvement as an informant, the role he played in the Stockheimer arrest, the laetrile connection, and the emergence of Representative Lewis. Like the telling of a movie plot, he introduced each of the major characters and events that led up to his trip to Madison.

They talked about possible charges and Tuerkheimer asked for the agent's own opinion. "After discussing it with Milwaukee, we thought it might fall within the boundaries of Neutrality Statutes," Burg said. "The fact is, though, we're not sure. We have something; we're just not sure what it is, and we need your help."

Tuerkheimer listened intently, sometimes impatient for certain details. "Is Lewis physically present, yet?" he asked. "Has he showed up at any of the meetings that Muckerheide has attended?"

Agent Burg told him about the meeting in Wausau.

Tuerkheimer wanted more. "We need irrefutable, physical evidence. We need pictures. We need an agent's eyewitness account. Bring that to me, and we have a case."

Tuerkheimer had one other request: he wanted a personal meeting with Mike. He needed to know more about his potential star witness: Did he have purple hair, dress like a hippy or act strange? Having people tell him that the scientist was a great guy wasn't enough: he had to learn it for himself.

Abruptly, Tuerkheimer ended the conversation and encouraged

---

[39] John B. Connally, Treasury Secretary in the Nixon Administration, faced five charges in a milk price-fixing case in 1974. Three of the counts were dismissed and Connally was found not guilty on the two remaining counts of accepting an illegal payment.

On the Laser's Edge...92

Burg to keep in touch. Burg promised an introduction with Mike as soon as possible, and left feeling good about the case, though still a little uneasy about Tuerkheimer. The man seemed pleasant enough. He wondered if he would still be feeling that way several months later; or, by then, would he also be calling him "that goddamn Tuerkheimer"?

The meeting was productive, albeit not quick enough. When Agent Burg returned to his Bureau car, it was missing. Unfamiliar with the strict, no-leniency policy of Madison parking rules, the agent had fully expected the parking police to leave him alone upon seeing his red gumball on the dashboard. He was wrong. Instead of towing the car miles away as they could have, however, the parking police showed a measure of mercy. They towed the car around the corner to a parking zone, "with a ticket attached to the window," Burg noted. To avoid attracting the attention of his employer, he got out his wallet, quietly paid a hefty fine at the municipal office across the street, and left town.

Mike was relieved to hear of Tuerkheimer's support. A few days later, he met with the prosecutor in Madison. It was a good meeting. He decided he could trust Tuerkheimer to help him. At long last, maybe his dark, secret life was finding some daylight.

Tuerkheimer had insisted on seeing more physical proof of Lewis' involvement with Iwen. It could not have been timed more perfectly. Another meeting was being planned for late June. The legislator would be on his way to Northern Wisconsin by car on a personal trip with his girlfriend. Claussen called the Holiday Inn in Stevens Point several days in advance to make reservations for a noon lunch.

Just as the conspirators laid plans for the meeting, so did the FBI. An extra agent was called in from Milwaukee and the second Wausau agent, Szekely, was assigned to assist.

On the day before, the three agents met at the Wausau office to outline their plans. There would be a total of three agents working the scene. Two agents, Szekely and the Milwaukee agent, would be inside the Holiday Inn having lunch, carrying with them a briefcase camera. Agent Burg was to be stationed outside in the parking lot, with a conventional, 35 mm camera.

Next day, six people showed up at the Holiday Inn: Iwen and his secretary, Lewis and his girlfriend, Mike, and Claussen.

Code Word Tikal...93

By plan, Mike was the last to arrive: the FBI had planned it that way. They wanted everyone else in place first before any conversation about lasers. Mike was instructed to park in a conspicuous spot so he could be photographed showing off the laser-burned items he had stashed in the back of his car.

Laetrile, politics and lasers occupied discussions at the meeting. There was another trip to Guatemala being scheduled for the fall, and an upcoming meeting in Florida with Fuentes. For both trips, the men insisted, Mike needed to go along.

Lewis and Iwen showed off letters they had received from Latin American officials. At least one was from Fuentes on government letterhead. Another was from the government of Paraguay. The politician and the chemist were enthusiastic about the progress they were making, but clearly concerned; they needed to come through with the goods to prove their ability to do so. Naturally, the responsibility fell on Mike. Lewis and Iwen both insisted that he had to make a laser weapon or the negotiations with foreign countries would come to nothing.

Lewis was still focused on an attack of the O'Hare Airport control tower to disrupt aircraft entry. Iwen's preference was still Jewish properties. "Iwen wanted to show Guatemalans that his intent was extremely serious," Mike remembered, "that he could produce this effect, get the publicity needed for Guatemala to read and say, 'See, Iwen is really doing his job.'"

At the meeting, Iwen said, "And you men can do that in the summer because it's going to be a hot summer and many politicians are going to be attacked."[40]

Iwen was candid, but he was not without his suspicions that day. He was sure that there were agents in the room, watching, listening. Mike made light of his concerns. "Yeah, they probably got this olive bugged." He picked up an olive from a relish tray and started talking into it as if it concealed a microphone.[41]

After lunch, Mike led Lewis and Iwen out to his car. From the open rear of the stationwagon, he pulled out a firebrick. A small,

---

[40] Grand Jury testimony of Myron C. Muckerheide, May 2, 1979. case #79-CR-57, *United States versus James R. Lewis*, U.S. District Court, Western District of Wisconsin.

[41] Transcript of U.S. Secret Service recording between Kevin Buggy and Albert C. Iwen, December 5, 1979.

On the Laser's Edge...94

smooth hole had been drilled through its center with the light of a carbon dioxide laser. A second brick was bluish-gray, hit with a pulsed laser, "bursting it as it impaled it." Yet another sample was a cylindrical piece of metal cut cleanly in half. "The atomic weapon at Hiroshima had turned brick to glass and that same affect can be produced by laser," he told them. Impressed, his audience agreed that the samples should be among the items taken to Guatemala to show government officials.

Throughout the meeting, inside and out, FBI agents were discreetly snapping away, capturing the moment for pictured posterity. Dismissive of Iwen's suspicions, the group never guessed that the two agents inside the dining room were anything but two businessmen having lunch. Outside, Agent Burg was not so lucky. When the group dispersed, he didn't notice that one of the women, an attractive, slender woman with a stream of dark hair, had gone back inside the Holiday Inn. When Lewis' girlfriend, Deborah, returned to the parking lot to leave with the assemblyman, she spotted Burg in his yellow, green vinyl-topped sedan aiming his camera towards their car.

Burg finally noticed her in his rearview mirror as she rushed to alert Lewis. He sped off through the parking lot before she could see his face.

Batzler and Lewis jumped into his Buick. Quickly, he jammed it into gear and sped after the agent's sedan. He got close enough to read its license plate number, JM1374, then left Stevens Point, headed north.

Lewis was so unnerved that he stopped repeatedly at Sheriff's Departments all along his route to run the license number: first at Wausau in Marathon County, then Merrill in Lincoln County, Rhinelander in Oneida County, and finally his destination of Eagle River in Vilas County. In 1977 license numbers were a matter of public record in Wisconsin, and anyone could request and receive the information; but, in the case of Bureau cars, each request kicked out a notification to the FBI, alerting them immediately when someone was checking up on one of their vehicles. Agents thought the Bureau used it to keep track of their own agents — what speeding violations they received, or where they were at a given time — but it could be useful as well in situations like this.

On each attempt to run the number, Lewis received the same information: the car was registered to Matt Q. Helm. The name was a dead giveaway. Helm was a fictional detective who had moved

Code Word Tikal...95

from books to the big screen for a popular series of spy spoof movies in the late 1960s and 1970s starring actor Dean Martin.

At the time, all the Bureau cars in Wisconsin were registered to fictional characters. In addition to Helm, cars were registered to Dick Tracy and Mickey Mouse. The addresses were fictional too. Commonly used was the 10,000 block of East Wisconsin Ave., Milwaukee, putting it somewhere in the middle of Lake Michigan. Matt Helm lived at the nonexistent 4567 Sweet Street in Franklin. Such creative licensing was understandable coming from the man responsible for the Bureau's licensing in Wisconsin at the time, an agent out of the Milwaukee office with the lyrical name, Friend Adams.

Lewis was desperate to know who the real Matt Helm was. What did he — or likely they — want? How much did they already know? He tried to answer those questions by calling the Chief of the State Capitol Police in Madison. Several times the agitated legislator called Bob Hamele, putting the pressure on him to discover the true identity of Matt Helm and the reasons behind the episode in Stevens Point. He also made a phone call to Iwen.

When he learned of the car and the pictures, Iwen must have felt his worst fears coming true. Everything he had worked to put together was close to unraveling. Hopefully, Lewis could use his connections to find out what was going on.

Iwen called an emergency meeting that evening with Mike at Radtke Park. His agitation clearly visible, he told Mike, "It's almost Apollo, but not quite."

The term Apollo had a special meaning to the Wausau trio. Iwen, Claussen and Mike had started using a system of codes for their meetings. They called it the Keystone Code, named after an investment fund familiar to Claussen, and its tongue-in-cheek reference to the Keystone Cops. The code used a different symbol for each of their favorite meeting sites, such as K-3 for the Ponderosa Restaurant, and S-3 for Shakey's Pizza: a total of twelve sites. Apollo, symbolized as a triangle, was the unlucky thirteenth. It had nothing to do with meeting sites; it meant that something had gone wrong with the plan and they would need to leave the country immediately.[42] In Greek mythology, Apollo was the son of Zeus, the god of light.

---

[42] Copy of Keystone Funds note card from Mike's files. Confirmed in articles in the *West Bend News* and *Wausau Daily Herald*.

On the Laser's Edge...96

Mike was not surprised by the phone call. He had been alerted about the blown surveillance by Agent Burg hours before. Now it was time to assess the damage.

After agreeing to meet Iwen, Mike made a quick call to Burg to let him know what was happening. "I have to meet Al in a few minutes at Radtke Park," he told Burg. "It doesn't sound good."

"I don't think you should meet with him tonight," Burg advised. "He's not likely in a good mood and it might be dangerous."

Mike didn't let on how scared he really was at meeting Iwen at a deserted park in the darkness of night. If he thought he had a choice, he wouldn't have agreed to go. But, failing to show up might open him up to suspicion and blow his cover.

"I already told him I'd be there," Mike countered. "He'll be more suspicious if I don't show. Too much could happen if it all falls apart now. I need more time to protect my family."

"I can't tell you what to do, Mike, I can only advise you, but be careful. If at all possible, try to get him some place where there's lots of people around."

"I'll suggest Mr. Steak's, near the park."

"I still don't think you should go, but if you do, good luck."

It was an evening, like many, when Mike was working late at the laboratory. He had not been home yet, and he knew that Pat deserved to be warned. At the same time, he didn't want to alarm her. "The FBI is really making some progress in this case," he assured her by phone. "The noose is tightening on these guys. But you need to be careful. Lock the doors and don't answer for anyone you don't know. I'll be home late."

Pat sensed that something was wrong. She checked the doors, kissed their daughter good night, and picked up her well-worn Bible to read.

Radtke Park was a small neighborhood park located about midway between Mike's house and his laboratory. Though located in town, it was an isolated area at the end of a residential street overlooking the Wisconsin River, butting up against a strip of no-man's land belonging to the Wausau Municipal Airport. As he entered the park, Mike paused a moment, breathed in deep, exhaled long, and ventured on.

Mike pulled into the park at around 9:00 PM. Only a small light illuminated the area. Iwen was already there, sitting in his Mercedes. Mike pulled alongside. "You know, Al," he said through the open window, "I think I was followed. Let's go to Mr. Steak's."

It was not difficult to convince the cautious chemist that someone might be lurking in the shadows, especially because of what had happened earlier in the day. They circled around and headed back to Mr. Steak Restaurant, located just a couple of blocks from the park on the corner of the main street.

As the two men entered the restaurant, Mike's heart dropped. They had hoped for a busy, crowded restaurant where they could talk without being overheard. What they got were two old men and dead quiet. Iwen motioned for them to leave.

"Get into my car," he told Mike bluntly.

At first, Mike didn't budge.

"Get in," he ordered.

Reluctantly, Mike complied. "What's this all about?" he asked, trying to conceal his fear with a voice of impatience that had become a part of his standard form. "What's going on?"

"There's been a leak. Someone was taking pictures of us at the meeting in Stevens Point this afternoon."

Mike felt nervous as they drove away, headed back to the park. He kept stealing glances into the side view mirror, convinced he saw the lights of someone tagging. Was it Agent Burg? He hoped not. If Iwen even suspected that someone was following them, he might bolt and run, taking Mike with him. Crazy things happen to a man who feels deceived, and Mike found no comfort in knowing that Iwen carried a loaded gun in his car. When they turned into the park, the lights behind them disappeared. Mike was relieved, until he realized that no one would be there to help him if things turned bad.

Sitting in the dim light of Radtke Park, Iwen began to explain about the photographer in the Matt Helm car. "Jim is very uptight about it," he said. "He says he's going to come down on this with all the power he has to get this person." There was a stern, evenness to his words and he was clearly disturbed; yet, calmly, deliberately, he leaned over and pulled out his gun from a console between the seats. "If Apollo comes," he said, "I'm headed for South America."

"Are you going to shoot your way there?" Mike tried to make light of the situation.

"If I have to. Tonight, I would have shot anyone who followed us."

As he held the weapon, Iwen looked at Mike with a strange intensity, his gaze driving like a nail through the thin light. "I think there's been a leak," he said, and menacingly jammed a clip into his gun. Mike's memory flashed back to a similar moment in time, less than a year previous, when Iwen confronted him about the Stockheimer arrest. He had been able to bluff his way through it that time, but this night his nerves were frayed.

With his free hand, Iwen reached towards the ignition key. Mike was convinced he planned to drive someplace even more secluded. He worried about Pat and little Susan and wondered if he'd ever see them again. The moment was suspended in space, surreal and timeless. Had it all come down to this? Not since West Virginia had Mike felt so helpless and scared.[43]

Suddenly, unexpectedly, Iwen took his hand away from the ignition key, wrapped the gun in an oily rag and put it way. He turned to Mike and said, "You know what I like about you, Mike? You've got guts."

The episode was just one more test, and once again, Mike had passed.

For the FBI, it was time for damage control. Agent Burg was receiving alerts about Lewis' search for the real Matt Helm, so he knew his car had to be taken out of circulation immediately. He parked it overnight inside the garage at his home in the Wausau suburb of Rothschild and the following day another agent drove it back to Milwaukee to exchange it for another.

The FBI also contacted Bob Hamele at the State Capitol and enlisted his support. Hamele was a graduate of the FBI National Academy at Quantico, Virginia so he understood the confidential nature of investigations. As well, he was a good friend with the senior agent of the Madison office, Henry Curran. It was Agent Curran who called Hamele.

"Bob, we need your help," Agent Curran started. "It's about Representative Lewis, an assemblyman from West Bend."

"So, there is something going on. I've suddenly been getting a lot of phone calls from Lewis. Is Matt Helm one of your guys?"

"I can't tell you a lot at this point, but Lewis is in the middle of something that we're working on."

"How can I help?"

---

[43] Incident reported in Part III of "Lasers, Laetrile: The Wausau Connection," *Wausau Daily Herald*.

Agent Curran suggested that Hamele tell an abbreviated truth. He could tell Lewis that the FBI had contacted him, then simply ask the legislator a question: was he associated with anyone involved with drugs? Hopefully, that would cause Lewis to jump to the conclusion that the incident in Stevens Point was part of Iwen's continuing headaches with the FDA over laetrile.

Hamele obliged his friend and contacted Representative Lewis. The plan worked: Lewis was comforted with the thought that Iwen was the intended target of the surveillance and it all centered on laetrile. Laetrile had less potential for political fallout than lasers, especially since Lewis' support of the drug was already well known. Now life for him could move forward. To the relief of the FBI, so could their investigation.

While his own trigger-happy finger on a shutter release had compromised his car, Agent Burg's identity remained concealed. No one knew for years, even within the Bureau, who was the real Matt Helm. "For a long time agents would ask me, 'Who is Matt Helm?' and I wouldn't tell them," said Agent Burg. "Then, one day, I finally told them, 'I'm Matt Helm,' and they laughed. They didn't believe me." The mystery had taken on a life of its own, no one willing to accept the connection of the fun-loving movie character to the conservative Burg.

As for the fictional Matt Helm, he disappeared from the Bureau ranks, never to return. After the surveillance fiasco in Stevens Point, the Wisconsin Bureau devised a less creative system for registering vehicles, bordering on the mundane.

But what they lost in creative licensing, the FBI gained in evidence. They had captured irrefutable proof that Representative Lewis was rubbing elbows with some troublesome, extreme, right wing conspirators. Finally, maybe more people would take Mike's Muckerheide's bizarre adventures more seriously.

On the Laser's Edge...100

In later years, Mike (right) came to know Dr. Theodore Maiman (center) and had the honor of seeing the first working laser, which Dr. Maiman is holding. It was Dr. Maiman's introduction of lasers on national TV that piqued Mike's interest and changed the course of his life. Dr. Dudley Johnson, one of Mike's closest friends, is pictured on the left.

Mike (front, right) meets with Melvin Laird and Caspar Weinberger at the University of Wisconsin-Stevens Point laser laboratory which Mike helped design. Dr. Ronald Uecker is standing to Mike's right.

Electrocardiogram Sent To Wausau From Thailand

Transmission of electrocardiogram data half-way around the world from Bangkok, Thailand, to Wausau by radio and telephone was completed Thursday night.

Myron Muckerheide, medical-electronics researcher at Wausau Clinic, devised the telephone transmission test which was conducted with the cooperation of Howard Gernetzke of WSAU-TV. Gernetzke is on a Far East tour with a group of area persons, and telephoned the tape recording of an electrocardiogram test from Bangkok at 8 a.m. Thursday. It was 8 p.m. Wausau time when the information was received but the actual time delay in transmission was less than a second.

Normally, electrocardiogram results are viewed on a tape. However, for the test, the tape information was converted to sounds which were transmitted by radio from Bangkok to Oakland, Calif., and from there to Wausau by radio microwave and telephone cable. The sounds are then converted back to a tape for analysis.

The background was noisy but the transmission will be fed through a computer to produce the electrocardiogram tape for study, Muckerheide said.

Except for finishing touches on the system, the test is proof that immediate consultation can be obtained between any two points on the earth, Muckerheide reported.

In the past, the problem has not been that doctors could not talk between Bangkok and Wausau, but that one doctor did not have the information necessary to make the diagnosis.

10-16-65

Article from the Oct. 16, 1965 *Wausau Daily Herald* about Mike's early successes in electrocardiogram transmissions.

In the 1970s, Mike was doing work in laser research which included its effects on aircraft.

On the Laser's Edge...102

Mike's Pinto stationwagon led a long and colorful life. Here, one of Mike's prototype laser devices is mounted on its top. Mike is pictured on the left.

The laser laboratory at Northcentral Technical Institute in Wausau, Wisconsin sometime in the 1970s. It was one of several laboratories Mike helped to design. He also did experimentation there.

Code Word Tikal...103

Rep. Jim Lewis (far left) and Al Iwen with their female guests arriving at the Stevens Point meeting which was secretly surveilled by FBI agents.

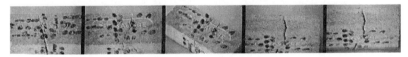

Laser pierced bricks were often carried by Mike to show people the impact of laser beams.

Mike's frumpy Pinto stationwagon in the parking lot at the Holiday Inn in Stevens Point, WI the day of the Matt Helm incident.

On the Laser's Edge...104

A more recent photo of the old Federal Office building in Madison, Wisconsin where Grand Jury testimony for the laser case was taken.

The hallway in the the old Federal Office building in Madison hasn't changed much, with its marble flooring and long oak benches.

Code Word Tikal...105

Mike was first introduced to Rep. Lewis at Radtke Park in Wausau, WI. It was also here that Iwen questioned Mike about the Matt Helm incident, when the FBI nearly blew its surveillance cover.

```
MONDAY SEPTEMBER 18, 1978
8:35 A.M.
VERY NECESSARY TO MAKE PRESENTATION
VERY GOOD FOR OUR PROJECT AS LONG AS IT IS COMPLETE
MUST BE A GOOD SHOW - TO CONVINCE OTHERS
VERY IMPORTANT
BETTER TO WAIT THAN PARTIAL SHOW  TWO MONTH DELAY O.K.
MIAMI: SKYTRONICS
       COL. DON EGGLESTONE
BROWNSVILLE  8 HOURS
```

Notes taken from a phone call between Rep. Lewis and Col. Fuentes, a Guatemalan government official.

On the Laser's Edge...106

SUBPOENA TO TESTIFY BEFORE GRAND JURY

United States District Court
FOR THE
WESTERN DISTRICT OF WISCONSIN

To Mr. Myron C. Muckerheide
    5510 Pine Park
    Schofield, Wisconsin

You are hereby commanded to appear in the United States District Court for the Western District of Wisconsin at Rm. #320, Grand Jury Room, 215 Monona Ave., Fed. Bldg., in the city of Madison, WI on the 25th day of October 1978 at 2:15 o'clock P.M. to testify before the Grand Jury and bring with you¹ passport; copies of all correspondence, documents and communications to or from any official in any foreign country; all correspondence among public officials related to Latin American affairs; all correspondence or communications with James R. Lewis, Albert C. Iwen, Debra A. Batzler, John F. Claussen, Joan Cimino, and David Pennings.

This subpoena is issued on application of the United States of America.

JOSEPH W. SKUPNIEWITZ
Clerk.

Date October 10, 19 78.
By _____
Deputy Clerk.

RETURN

Mike's subpoena to appear before the Grand Jury.

KEYSTONE FUNDS

B-1, HOFFMAN HSE.   MR STEAK (WEST) - B-3
B-2, NINO'S   SAMBO'S   - S-5
B-4, HOWARD JOHNSONS   PONDEROSA   - K-3
K-1, AIRPORT N   APOLLO, Δ
K-2, DALES
S-1- PANCAKE HSE
S-2- MASTERPIECE
S-3- SHAKEYS
S-4- MR STEAK SCH.

Used as evidence was a card revealing the Keystone Code. The code indicated meeting sites between Iwen, Claussen and Muckerheide. The final code, Apollo, indicated major problems.

Code Word Tikal...107

The first series of articles on the laser conspiracy began to appear in the *West Bend News*, West Bend, WI.

Used as evidence in Grand Jury testimony was this letter from Iwen helping to introduce Mike to John Hartl in West Virginia.

On the Laser's Edge...108

Wausau scientist was key FBI informer in plot

By HERALD STAFF
(EDITOR'S NOTE: The accompanying article, published today by the Wisconsin Associated Press, gives a broad outline of a story that has been researched by the Herald for the past four weeks. Our research began when Schofield laser consultant Myron Muckerheide volunteered to tell us about his experiences as an FBI informant. Reporter Pat Rupinski followed up by checking with other sources and looking at court records. In one court file, relating to the perjury conviction of former state Assemblyman James R. Lewis of West Bend, the Herald found documents and other information that had been presented to a grand jury investigating the laser case. Today we begin a five-part series, detailing the account as it has been told to us and from what we could learn from court files and other sources.)

On April 1, 1976, Myron "Mike" Muckerheide went to see presidential candidate George Wallace at Central Wisconsin Airport, Mosinee. He was accompanied by John Claussen of rural Wausau, who introduced Muckerheide to Albert Carl Iwen, then of Merrill.

Iwen was chairman of the American Party in Lincoln County in 1972. That same year, he unsuccessfully ran for mayor of Merrill and for the state Senate in the 12th District. Later, he lost elections for Merrill alderman in the city's Third Ward and for a seat on the Merrill School Board.

According to Herald records, Iwen was once an officer of Mosinee Research Corp., a now closed firm also known at times as U.S. Pharmaceuticals Inc.

In April 1976, Iwen was known publicly as a local businessman and an unsuccessful conservative political candidate.

At their first meeting, Muckerheide says, Iwen talked about lasers. Muckerheide didn't find that unusual. He says many people ask about his work.

But he recalls some unusual overtones in Iwen's conversations.

In a recent interview with the Herald, Muckerheide said he was asked if he knew anything about force fields. Iwen said he knew of a foreign country that was extremely interested, Muckerheide added. He said Iwen told him there was "plenty of money available."

Muckerheide says the conversation aroused his curiosity. The following day, he went to a plant Iwen was affiliated with, Mosinee Research, to see him.

Muckerheide says they chatted and then Iwen showed him his "laboratory."

"I was later to learn it (a plane with a laser weapon) would be flown to attack Cuba."
—Myron Muckerheide

Muckerheide recalls: "He (Iwen) took me into the back room, and there were men working back there and the place smelled like alcohol or ether. There were large magnetic stirring devices operating and there were men working very vigorously.

"The windows were opened, I remember, and they were making something to do with amygdalin."

Amygdalin is a white crystalline glucoside that occurs in the kernels of fruits such as almonds, apricots, cherries and peaches. The amygdalin obtained from apricot kernels is used in the manufacture of the drug Laetrile.

Whether the Mosinee plant was actually making Laetrile may never be known. The plant burned in the fall of 1976.

(After the fire, Mosinee Research moved its operations to a former dairy building in Manitowoc. On April 12, 1977, Iwen and Douglas Evers, the Manitowoc plant's manager, were arrested when they obstructed FDA agents who wanted to search the plant. The agents got a federal search warrant. The FDA then seized some of the plant's items and analyzed them. Some of the items were believed to be used in manufacturing Laetrile and the plant was ordered to close.)

The FDA raid was Iwen's first known trouble with the law. However, the FBI's interest in Iwen started after his introduction with Muckerheide.

After talking with Iwen, Muckerheide contacted the Wausau office of the FBI and reported his con-

See WAUSAU, Page 12, Column 3.

The Wausau Daily Herald (left) and *Detroit Daily News* (below) each featured a series of articles about the laser case after the court files were made public.

The Detroit News
MICHIGAN'S LARGEST NEWSPAPER

March 8, 1981
108th YEAR NO. 198
50 CENTS

He tells of plot to attack O'Hare Airport, Havana

By HUGH McCANN
News Staff Writer

WAUSAU, Wis. — For three years, Myron C. Muckerheide walked between the law and conspiracy, a party to what he says was a plot to incite nuclear war and revolution while he secretly passed information to the FBI.

The double life preyed on his mind. At times, the 50-year-old laser scientist thought he was mad to believe that here, in his hometown, acquaintances were plotting violent revolution; that a superweapon he had been asked to build was the linchpin in a plan to destroy targets such as Havana harbor and the control tower at Chicago's O'Hare Airport. At least twice in the course of clandestine meetings, Muckerheide was sure he was going to be killed.

THE FEAR THAT built in his mind is still there. Now, he is bitter as well.

He is bitter that, although he told his bizarre story to a federal grand jury, no charges resulted,

no trial was held. And those he tracked for the FBI are now free men, as far as he is concerned.

Men who "desired harm for my wife, my child, your wife, your children" are today "walking around free."

Muckerheide is bitter as well with the American press, which stood by and let a charade of justice take place, as he sees it.

Muckerheide's story has all the elements of a science-fiction potboiler.

The cast of characters includes a Wisconsin politician, an investment counselor and a pharmaceutical manufacturer who used titles borrowed from Nazi Germany. It involves at least two Third World nations and an armed American right-wing brotherhood.

MUCKERHEIDE SAYS he stumbled onto the group in 1976 and went to the FBI. The FBI asked him to play along, providing him with a tiny tape recorder he used to record secret meetings. He used such devices as laser-burned bricks to help convince the group that he was making progress on a laser weapon.

After three years, Muckerheide told the Wisconsin grand jury in Madison, Wis., that he had uncovered a plot to:

● Establish a secret laser weapons laboratory somewhere in Guatemala.

● Equip an aircraft with a powerful laser gun and zap Havana in hopes of triggering a Russian nuclear counterstrike on the United States.

The Muckerheide Intrigue

First in a series.

After that, Muckerheide told The Detroit News that an extremist army would be unleashed into the postwar rubble of America to set up the Second Republic.

THE INVESTMENT counselor, John F. Claussen, said in a sworn statement to the FBI that the group wanted a laser weapon

Continued on Page 4A

Code Word Tikal...109

Mike prepares his first NASA Get-Away Special (GAS) payload for launch.

Mike speaking at the Goddard Space Flight Center.

On the Laser's Edge...110

Mike standing in front of the space shuttle exhibit at the National Air & Space Museum in Washington, D.C.

Shuttle Exhibit — Milwaukee's Contribution

In 1986 St. Mary's of Milwaukee was the first hospital to design a series of tests for the Space Shuttle. Under the name PROJECT JULIE (Joint Utilization of Laser Integrated Experiments) twenty different trials were completed in weightless outer space. Their story and the hardware from that significant project is on display.

A display at Mitchell Airport, Milwaukee, highlights Mike's first NASA GAS payload, Project Julie, for St. Mary's Hospital.

Code Word Tikal...111

Mike in later years, speaking to a group of laser scientists at a Minneapolis convention.

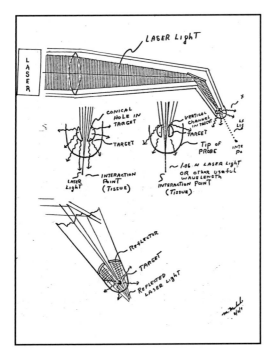

One of Mike's work drawings.

CHAPTER FOURTEEN

Politics is the art of looking for trouble, finding it whether it exists or not, diagnosing it incorrectly, and applying the wrong remedy - Ernest Benn

Sometimes it takes a different perspective to make a plan work. The plan to install a laser weapon into an airplane and blind air traffic controllers was certainly more plausible than today's random efforts at blinding pilots, (it is easier to hit a stationary target than a moving target); but, it was just too ambitious for the times and Iwen's small, meagerly financed group. It took Lewis' perspective, Mike's orchestrated maneuvers, and Iwen's continuing saga with the FDA and laetrile, for him to get reasonable.

By the time Iwen returned from his second trip to Central America, laetrile was well on its way to taking a back seat to lasers. Lewis had convinced him that, even though Guatemala allowed laetrile production, the country's friendly ties to the U.S., especially concerning health regulations, might cause problems. The weapons industry, however, presented an untapped opportunity. There were always wars someplace in the world, promising an eternal fountain of profit. At the same time, the financial freedom would give him the opportunity to develop what he really wanted: something to instigate his own war.

It was still being assumed that the lasers would be made in the U.S. and shipped illegally out of the country. The acceleration of Iwen's desire to accomplish his goals, however, started bothering Agent Burg. It was clear to him that Mike had to start back pedaling and downsizing these grandiose plans.

On the one hand, the weapon size had to be shrunk to fit within the law of what the U.S. Attorney could prosecute. On the other hand, Mike could never be allowed to produce the weapon because of the possible hazards and legal implications it posed. Somewhere in the mix was the believability factor: what judge, what jury would ever believe that a couple of independent, no-name scientists from the middle of Wisconsin had the ability to install an effective laser weapon into the nose of an airplane? Consequently, Mike started advocating for a smaller weapon.

Iwen bought into part of the downsizing. A smaller weapon would be easier to smuggle out of the country, he reasoned. Better yet, maybe they should move the entire manufacturing process out of the United States to avoid the complications of shipping. His contacts with Lewis and Fuentes assured him that his future lay in Guatemala. As with his laetrile plans, not yet forgotten, he decided he would set up a factory in the *Land of Eternal Spring* to make lasers. Handheld weapons would pave the way for the bigger devices he especially sought. He offered to give Mike 51% of company ownership in exchange for laser technology. When Mike asked what kind of market he anticipated, Iwen told him, "There's a lot of unrest in Central America right now. A lot of potential for sales." He talked about meeting with representatives of other countries who were also interested in high-powered lasers. "When we were down there," he added, "the colonel told me that he intended to supply help to Somoza in Nicaragua."

By early June, Iwen's plan was already beginning to take its final, practical form. The earliest details of it were offered in a meeting two days after the Matt Helm incident, when Representative Lewis and his girlfriend were headed back to Madison. The laser group boldly returned to the Holiday Inn, suspiciously surveying the room for potential spies. They noted a couple of possibilities — some businessmen having lunch — but they were wrong. The FBI had decided not to send any surveillance agents out for the meeting, and the men in question were just men having lunch.

Because the party included two laetrile supporters who knew little or nothing about the laser plot, it was only in the parking lot that the subject of lasers was brought up between the four insiders. Iwen was advocating that Mike build a laser in the U.S. to be sold to Guatemala. "Once they see we can produce the product it'll buy our way into setting up a bigger production plant inside Guatemala," Iwen believed. "There, we can make the big stuff."

Lewis motioned his head towards his late model, four-door Buick sedan and said, "Yeah, if I had a laser weapon in Guatemala, I wouldn't be driving this thing around. I'd have an army at my disposal, and drive a Mercedes. I'd have my own airplane, and a ranch. I could live there, easily. Debbie has arthritis. Guatemala would be an excellent place for her."

In the real world, Iwen wasn't far off the mark. In the dark, impenetrable forests of military laboratories, blinding laser weapons existed in 1978. Even today, however, no one freely admits to when they were first used and by whom. Although the U.S. is thought to have used lasers to blind the optics of Soviet-made North Vietnamese antiaircraft systems,[44] reports claim its use to blind soldiers was used by the Vietnamese using Soviet-made weapons in a 1979 border war between China and Vietnam. More reports of blindings sifted out during the Soviet conflict with Afghanistan in the 1980s when a surprising number of blind people started showing up in hospitals. Since then, there has been periodic speculation about the covert use of ocular-aimed laser weapons, surfacing and then submerging again in the media, like sightings of a Lock Ness Monster.

To people in the know, then, Iwen was beginning to make sense. With daily confrontations of warfare in several Latin American countries, governments and civilian guerrilla groups were hungry for guns and ammunition. They often acquired them through legal and illegal means. In countries where "elected" dictators ruled, and entire villages were massacred on whim by both sides of the battle, it was not a stretch to imagine that Iwen would eventually find someone who would buy the laser technology he thought he had to sell, and with Lewis' influence, he was getting closer to finding them.

So far, Mike had not done much to help Iwen accomplish his goals. He frequently carried in his car his laser-struck samples, but the chemist would not be forever happy just looking at what a laser could do. He needed a laser weapon to show to potential buyers and backers and he started pressing Mike to deliver one.

Agent Burg knew that he could never let Mike do that; a laser weapon might end up in the wrong hands, smuggled out of the country, and broken down and analyzed. The scenario reminded him of the story he'd heard about an FBI case in Pennsylvania. The Bureau had been tipped about the planned theft of a shipment of furs. The agent in charge directed that a trailer be loaded up with furs, calculating that

---

[44] *Laser Weapons: the Dawn of a New Military Age* by Major General Bengt Anderberg and Dr. Myron L. Wolbarsht, 1992 Plenum Press.

agents laying in wait would pounce on the thieves before the load was stolen. Unfortunately, something went wrong, and the entire trailer load disappeared before the FBI could catch the thieves. The agent was demoted and transferred to another office where other agents familiar with the story would call out to him as he passed their offices, "Load 'em up!" From that incident, Agent Burg learned: "You manage the job in the safest way possible. You don't let them *load 'em up.*"

"Instead of a laser," Agent Burg approached Mike, "what about a film: something to show people that you know what you're talking about without actually giving them anything?"

Mike thought it might work and approached Iwen with the idea. He pointed out that it was much too risky to show a buyer an actual laser weapon. "After they have the weapon, they have the technology," he argued. "After that, I'm disposable. I might be eliminated, or in the least, cut out of the profit."

Iwen agreed; after all, he too could be cut out of the money once a product was delivered. He gave Mike instructions to make the film instead.

Mike had several films that showed a laser at work and could have used any one of them from his available collection. That, however, would have been too simple and too quick. He developed a string of excuses for not providing the film, from lack of equipment, to lack of time, to paranoia.

He would not be so lackadaisical with the FBI. Several agents came from the Milwaukee office to see the film, setting up a VCR in one of the rooms at the local Howard Johnson Motel. The film Mike chose was a carbon dioxide laser exploding air into plasma from one of his projects in Canada. The FBI found nothing that should prevent him from showing it to Lewis and Iwen; but, they encouraged him to keep coming up with excuses: if he could delay showing the film to Iwen and Lewis, it would also delay the showing of it to Fuentes, and hopefully buy the FBI more time to gather more evidence.

That task wasn't difficult to accomplish. Each of the men led busy lives and scheduling conflicts were a frequent, understood complication. Yet another good excuse was the lack of equipment available for public use. The early VCR, with its large, clunky tapes, was still an expensive novelty and there were few places in the Wausau area where a player was readily available.

On the Laser's Edge...116

The first chance for viewing the tape came in late July when the men scheduled a private showing at the local historical museum. Still hoping to buy a little more time, Mike said he saw a yellow car circling the museum as he entered. "You don't suppose it's that Matt Helm car, do you?" he asked.[45] Mike's cohorts may not have been able to understand his lack of speed, but they understood his paranoia; they all suffered from it. With the seeds of doubt planted, they canceled the show.

It was a busy and difficult summer for Mike. He would meet every few days with Iwen and Claussen, then report the results to Agent Burg. To escape the office, Burg would often suggest that they meet at a local park, the two well-dressed men glaringly obvious in the backdrop of pine trees and playground equipment. "I often wondered what people thought to see two grown men in suits sitting together on a park bench in the middle of the day," Burg reflected.

Although Claussen was still in the inner circle, he didn't seem to be providing anything substantial to the laser plan beyond an occasional remark or opinion. He often excused himself early to get home to his wife. Either he was a thrill-seeking spectator, the FBI speculated, or he was a man on a secret mission.

The meetings were a comfort to Mike, and the parks a welcome break from the glare of laboratory lights, professional scrutiny and political extremists. He saw Agent Burg as someone who really cared about his fate. "He didn't hesitate to move forward," Mike said. "He was not desk oriented and he knew there was a world out there. Some of these agents are stuffed away in a box and don't see the world. Tom knew. I felt the safest I ever felt with any of them, with Tom."

It was a good feeling to have at a time when the case itself was starting to slip out of control.

---

[45] Interviews with Agent Thomas Burg.

CHAPTER FIFTEEN

*Gold cannot always get you good soldiers,
but good soldiers can always get you gold.*
— *Niccolo Machiavelli*

As with any great plan, its success or failure frequently comes down to one thing: money; no money, no plan. So far, Iwen had not come up with enough of it.

Although Colonel Fuentes had showed promise, it was apparent that he wasn't going to help finance anything until he was sure that Iwen and Lewis could come up with a working laser weapon. It didn't help that Mike was always jumping his estimates for what developing a successful prototype would cost.

With his own finances dwindling now that prosecution was crippling his once lucrative laetrile operation, Iwen hopscotched around Wisconsin looking for money, still using laetrile as his selling point. Sometimes, Lewis joined him on these fishing expeditions, once flying by private charter to the northern part of the state to approach a wealthy retiree whose wife had recently died of cancer.

The men were not short on ideas. Mike reported to Burg that Lewis was in favor of using Mafia money if necessary: "And once they have Fuentes on board, they think they can support both the governments of Paraguay and Guatemala with laser weapons."

Despite the financial setbacks, Iwen was invigorated by the support of Representative Lewis. Already, the politician's pull had helped open doors, and it was obvious that his clout was enticing a foreign military leader into buying into the plan. There was regular contact now between Lewis and Fuentes.

By midsummer, the money situation was starting to improve and Iwen was getting anxious to return to Guatemala. Always trying to get a fix on where the money was coming from, Mike mentioned the expenses involved with the trip. "I told Iwen that I thought things were getting tight, with all the legal fees," he said. "But he told me not to worry. Later on we went out to his car. He opened the trunk and took out this black attaché case, flipped open the lid and here's all this money: stacks of hundreds and twenty dollar bills.[46] He didn't say where it came from, only that he could get more if he had to."

The money and connections were finally coming together. Iwen and Lewis started making travel plans, and this time they included Mike.

"Al told me I'd have an excellent job in Guatemala," Mike said. "I'd be able to afford anything I wanted." Colonel Fuentes supposedly had a place for him to work, "a secret site where I could work with the military." When Mike told Iwen he wasn't sure he wanted to leave Wisconsin, he was assured that, "once I saw the country, I'd see what a great place it was."

To help pave the way for a smooth journey, Representative Lewis wrote a letter to Senator William Proxmire complaining about the treatment Iwen had received by U.S. Customs on a previous trip. He may have been referring to a shipment of laetrile supplies from overseas that had been confiscated by FDA officials on its arrival to the U.S. in July. The sidetracked shipment had been a financial loss for both Iwen and Lewis.[47] In writing his letter of complaint Lewis could only hope for more cooperation on their next travel adventure south when they planned to transport the laser film, and possibly some hardware.

Of course, Lewis couldn't tell people that he was going to Guatemala to try to sell laser weapons illegally. His interest in photography and the photographic ruins of Tikal provided him the perfect excuse he needed.

"Jim was going to use the cover that he wanted to photograph the Mayan culture at Tikal and I was told I could use the excuse that I was just going along for a sightseeing tour," Mike explained. When he asked how much the trip would cost, Iwen told him he had paid for Lewis' trip in the spring and it had cost around $3,000. "I told him I couldn't afford that," Mike said, "So he offered to pay for it himself."

For the trip, Mike was instructed to get his materials together: the laser film no one had yet seen, and the samples of brick. The men had not convinced Mike to take along a prototype, but they had not given up trying.

Naturally, Mike didn't want to go, but he had to temper his

---

[46] Mike Muckerheide interviews.

[47] In an FBI taping between Representative Lewis and Mike Muckerheide, October 6, 1978, Lewis claimed that he lost $20,000 and Iwen $17,000 in the confiscated shipment.

protest. When August was suggested, he insisted that he couldn't possibly find the time. Then there were all the little details one must tend to in order to smuggle something out of the country that might arouse suspicion from Customs: like firebricks and laser films and maybe a laser. Did Guatemala have the kind of VCR equipment they needed to show the film? If they took a VCR with them, what kind of electrical current was used in Guatemala? And how would they get the items in and out of the country undetected? "That letter Jim wrote to Proxmire isn't good enough for me," Mike complained to Iwen. "I have a lot at stake here. How are we going to get something like that past Customs?"

Iwen had an answer. "I can find a church minister whose willing to collect relief supplies for Nicaragua and Guatemala, a friend of mine. And I can help with that." He said the laser items could be smuggled inside the relief supply.

"I don't like that idea," Mike continued to argue. "It could get lost in shipment. It could end up in a big container, buried some place. You might never find it again."

Iwen dismissed his concerns. "I know a pilot who can help us," he maintained. He doggedly pursued avenues for the relief shipment, looking among his extremist buddies for support. Several of them were lay ministers for small congregations who could provide a front for their humanitarian cover.

By mid-September, communications with Fuentes began to pick up and things were looking good with Guatemala.

"Jim has talked to the colonel by phone," Iwen reported, "And he says he's in urgent need of something like a laser weapon." He added that the Guatemalan official now had the backing of his government and was under a great deal of pressure to come up with a laser. It was news the Wisconsin men — sans Mike — had been waiting for. They started finalizing their travel plans.

A meeting was scheduled for noon on September 22 at the Howard Johnson's in Wausau. Iwen was late to the meeting with Mike, Lewis, Claussen and the two women, Batzler and Joan C. He seemed flustered by his delay.

The conversation settled into the pending trip to Guatemala. Although Mike had agreed to go along, he told the men that he wanted more proof that the Guatemalans were really on board with the plan. Iwen, who was particularly out-of-sorts that day, looked at

On the Laser's Edge...120

Lewis and asked impatiently, "Can I give this to Mike?"

"Sure," Lewis replied, and Iwen handed Mike two photocopied documents. Mike noticed that one was a letter signed *Fred Fuentes*; the second was hand scribbled notes. With little acknowledgment, he tucked them both into his briefcase.[48]

"What about the trip to Florida you guys were planning?" Mike continued to prod.

"It's been delayed again. Fred is a busy man, but he's really interested in what we have," Iwen said. "And if he can't make a trip to the U.S., I think we need to go Guatemala to meet with him as soon as possible. I'm looking at the middle of October, three weeks from now."

"That means we need to see the film soon," Lewis inserted. "If I can fit it into my schedule, I'd like to see it this coming Monday."

"I'll see what I can do. It's been really busy at the lab, but I'll try," Mike promised.

Afterwards, in the parking lot, Lewis took a cardboard box out of his car and handed it to Mike. "Here," he said. "You're going to be needing this."

Mike took the sealed package and put it in his car. He knew what was inside the box, and he knew where he was headed next. When he was safely away from the meeting, he found a pay phone and called Agent Burg. They agreed to meet at Marathon Park. There, in a heavily wooded area, Mike handed over the two pieces of paper and the cardboard box.

"These guys are getting pretty serious," he told the agent. "There's a gun in there."

Agent Burg took the sealed box and placed it inside the trunk of his car. With Mike looking on, he carefully cut through the tape and opened the lid. Inside the box, he found a Colt .38 revolver loaded with six rounds of ammunition and resting in a leather holster.

"These guys are all packing guns, Tom," Mike said. "What happens now?"

Agent Burg could tell that Mike was feeling the pressure. Things were getting serious. He had to make sure that Mike stayed calm.

"Mike, they're just burying themselves deeper," he assured him confidently. "They're just giving us more evidence to work with. This

---

48 Grand Jury Testimony of Mike Muckerheide, May 2, 1979.

is great!"

What Mike had seen as a hazard, Burg had seen as an opportunity, and he knew that the agent was right. "So, what do I do with this stuff?" he asked.

"Let me make copies of the letters and I'll return the originals to you in case they ask for them back. As for the gun, let me take pictures, then we'll get a safety deposit box at a bank and keep it there. We'll rent a separate box for it. Only you and I will have access to it. It'll help us keep a clean line between when it went in and when it went out. As for the cardboard box and the other packing materials, I'll take it in for fingerprint evidence."

Agent Burg instructed Mike to meet him at a local bank later that afternoon. In between times, he made the necessary photocopies and pictures. He also took a closer look at the two papers.

The letter was dated September 6, 1978 and addressed to Lewis. It read:

> Dear James:
> Situation about our studies about "Tikal" has improved recently.
> I would like to know when will be possible for you to come with all your cameras and equipment.
> I'm very interested in our project. Please contac [sic] Mr. Iwen. Send him my regards and tell him that I want more details in order to plan a show "Tikal Show"
> Sincerely,
> Fred Fuentes[49]

From previous conversations with Mike, the agent knew that the term *Tikal Show* was being used as a cover for the laser film.

The second piece of paper, the hand scribbled notes, was dated September 18, 1978 and read:

> Very necessary to make presentation.
> Very good for our project as long as it is complete.
> Must be a good show — to convince others.
> Very important.

---

[49] Government Exhibit #1, Lewis exhibit #7, case #79-CR-57, *United States versus James R. Lewis*, U.S. District Court, Western District of Wisconsin.

On the Laser's Edge...122

Better to wait than partial show. Two month delay o.k.
Miami: Skytronics
*   Col. Don Egglestone*
Brownsville   8 hours[50]

Burg wanted to know more about Colonel Don Egglestone and his connection to Colonel Fuentes. He began making phone calls and discovered that Egglestone was a retired U.S. Air Force colonel who operated Skytronics, Inc. out of Florida. The small company manufactured aircraft parts and was the sole distributor for certain aircraft-related items. About half of their business was as a freight forwarder and broker for military purchases by Guatemalan armed forces, or as the purchasing agent for the Guatemalan Air Force. The Guatemalan Air Force sent a DC-6 aircraft to Miami to pick up freight on a regular basis. Fuentes would occasionally make the trip to oversee the operation. There was no indication that it was anything except a legitimate business operation. Burg wondered if he would ever learn if it was anything else.

Burg made two copies each of the letter and the page of notations. One copy would go into the official case file in Milwaukee and one into his own file. His file was kept in the walk-in vault at the Wausau office. The vault was not standard for resident agency design. The Clerk of the U.S. District Court, the office's original occupant, had installed it. It was popular with the agents, who rejected offers to move to larger, more convenient quarters because they liked having a walk-in filing cabinet they would never likely get anywhere else. It kept its popularity despite its little quirk of never opening properly. After dialing the numbered combination a quick *thwack* of a mallet was needed to drop the recalcitrant combination tumblers into place. Agent Jack Page had left behind a complimentary mallet dubbed the *Jack Page Memorial Hammer*.

When the papers were safely filed, Burg turned his attention to the gun. He unloaded it and put the bullets in their own envelope, then prepared a special set of forms, or *green sheets*, that would follow the chain of custody. He made notations of the gun's make, model and serial numbers, took photographs, then started making phone calls.

---

[50] Lewis Exhibit #8, case #79-CR-57, *United States versus James R. Lewis*, U.S. District Court, Western District of Wisconsin.

He discovered that Lewis had arranged for the Colt's purchase at a Milwaukee gun shop just a few days earlier. He had picked it up two days before the meeting.

Although illegal while loaded, there was nothing illegal about possession of the gun itself; it only implied how serious Lewis was taking the situation. He wasn't the only one. John Claussen, who had not risen much in his role as a conspirator, showed off a nine-millimeter automatic pistol that he had started carrying in his car.

Iwen's final target date for departure to Guatemala was October 13. The trip would include himself, Lewis and Mike, as well as his attorney John Couture, and David P. who was also a lay minister. David P. and Couture were supposed to go to Nicaragua with relief supplies that were being collected as a front, while Iwen, Lewis and Mike went to Guatemala. In a phone conversation between Lewis and Fuentes the Guatemalan had again encouraged him to "bring the film."

"I want to go back out to Colorado to see Roberts before I go," Iwen said.

"What for?" Mike asked.

"In case we need him. If something happens, it's good to have someone who can help. He has connections. Can you give me something to take out to show him? Maybe another brick. I want to show him some laser stuff."

"Yeah. I can do that."

But on October 2, the weather turned and Iwen, who had intended to fly his own plane, postponed his travel. He would reschedule for the following week.

"Why don't you come with me when I go?" he encouraged Mike.

"I can't. I can't take a lot of time off now with the trip to Guatemala coming up. Too many things going on," he begged off.

In reality, Mike was feeling increasingly uneasy around Iwen. It was difficult to trust a man who wanted to blow up buildings and blind people and who saw himself and his plans as the savior of the world. Once, when the chemist offered him an apple, Mike's thoughts rolled back to the poisonous apple in the story of *Snow White and the Seven Dwarfs*. He declined.

Mike had more than a few things going on anyway, and the FBI had inspired one of them. Because Iwen had contacted a travel agency in Wausau and was working out the details for the pending trip, Agent

On the Laser's Edge...124

Burg had to find a way to keep Mike grounded on American soil. Allowing him to leave opened up too many dangerous possibilities. If he went to Guatemala, would he be allowed to return home? He had what many governments wanted — cutting edge technology in weapons development. Would they use him, then discard him, or keep him against his will? If Mike cracked under the pressure, and revealed his FBI connections, especially when out of the country, what would they do to him? Guatemala's recent history of abductions and disappearances was not a good omen. Agent Burg knew that, aside from any laser hardware that might be taken out of the country, they dared not risk losing Mike and his knowledge. They needed to find a way to stop the trip, and their best hunch was using Lewis.

Agent Burg realized that the best evidence was a confession made knowingly or unknowingly by the guilty party. If he could get Lewis to admit in his own words that he was trying to use his political clout to develop an illegal weapons sale to a foreign government, then the FBI stood a much better chance of developing a prosecutable case. Mike's ongoing reports certainly indicated guilt, but to have that admission on tape, in Lewis' voice, was the golden goose of proof.

Burg had seldom used a body wire to obtain a criminal confession, largely because the technology was still being refined in the 1970s, and because it was a complicated process requiring a series of official steps and documentation: another result of Edward Levi's 1976 surveillance guidelines. There had to be just cause for the taping, and at least one party involved had to give consent. In this instance, FBI Headquarters in Washington also had to be consulted because of the potential for political fallout in taping a public official. Regardless of all the red tape, it seemed like a logical way to go.

Burg first consulted with U.S Attorney Tuerkheimer to discuss if the taping would be of any benefit and whether Tuerkheimer would even allow it as evidence. Tuerkheimer approved the idea, and the agent took that backing to his supervisors in Milwaukee to ask for their opinion. Both Supervisor Byrne and SAC Hogan agreed that going the extra mile would be worth the extra risk.

Despite mutual agreement, the taping would hinge on one man's consent to bring it all together: Mike's. He was the only one they could rely on to wear a body wire and to get the evidence. If Mike rejected the idea, there was little for the Bureau to do but stand back and watch the plot take another step forward.

On one of Mike's routine trips to Burg's office, the agent approached the scientist with the idea. "Mike, we'd like to get a taped conversation between you and Lewis," he started. "Tuerkheimer agrees it would help us build a stronger case."

"You mean a wire tap on the phone?"

"No. We were thinking of sending you in with a body wire. We'd set everything up. It's not without risk, but it's generally a safe procedure. The taping would take place in a safe location. We're hoping you can set up an appointment with Lewis in his office at the State Capitol. What do think?"

Mike was weary from two-and-a-half years of cat-and-mouse games, but he understood that getting a taped confession from Lewis could provide new and valuable ammunition in the case. And, risking his life in Wisconsin was certainly better than risking it in Guatemala where no one was around to protect him. He was quick with his decision. "If you think it will help get these guys, I don't have a problem with it."

"Then you agree to do it?"

"Sure, I'll do it. When were you thinking?"

"The sooner the better. It won't take me long to get the written approvals. Let's try for this Friday. Do you think you could fit that into your schedule?"

"I'll make it fit."

On October 3 Mike called Lewis to set up the appointment.

"I have to be in Madison this Friday on business," he told the assemblyman. "I thought I might drop by and talk to you if you have the time."

"Sure. Anything in particular?"

"I just think with this trip to Guatemala coming up we should go over a few things," Mike said. To add some fuel to the fire, he added, "I can tell you now, though, I don't like Al's idea about the relief supplies."

"Why not?"

"I think it sets the trip up for too much publicity. And what happens if we try to smuggle the film in and it gets lost or Customs finds it?"

"I don't think we have to worry about Customs, but we can talk about it. Al never went to Colorado, did he?"

"No. The weather was bad. He's going next week instead."

"My secretary can schedule a time for us to meet. Friday, October 6, right?"

On the Laser's Edge...126

"Right. I have a 2:00 appointment at the university, so before that, if possible. Maybe 1:00."

"Hold on. Talk to my secretary. See you Friday, Mike."

That evening, the phone lines buzzed between Lewis in Madison and Iwen in Merrill. Lewis was relaying Mike's concern about the relief shipment. The next day, Iwen assured the laser scientist that he knew people willing to help get the shipment into a Latin American country safely: a pilot and a minister sympathetic to their cause. As for the conspirators, they would travel by commercial airline and he had contacted a local travel agency to work out the details for the October 13 flight. In the meantime, he would head off to Colorado to see Roberts and invite him along for the trip.

Mike buried himself even deeper than normal into his work that week, trying to forget everything that was spinning out of his control. It wasn't easy. He knew that if the secret taping with Lewis didn't work out, something else would have to be done quickly or he'd be off to Guatemala the following week, perhaps never to return.

CHAPTER SIXTEEN

*If you reveal your secrets to the wind,
you should not blame the wind for revealing it to the trees.*
— Kahlil Gibran

It was less than 24 hours from the secret taping in the state capital and Mike was meeting with Iwen. The chemist knew about the meeting in Madison the next day and he wanted Mike to deliver a list to Lewis. He didn't have it with him, but would make sure that Mike received it before his departure. It was a list exposing the identities of several CIA agents working in Taiwan.

Mike couldn't help but look perplexed. "CIA agents? What's that for?" he asked.

Iwen explained that Lewis was scheduled to go to Chicago the following week to meet with several Taiwanese officials. He proposed that the Wisconsin legislator give the list to the officials. It would curry favor with the Taiwanese government and possibly garner more support, financial or otherwise, for the group's development of laser weapons.

Mike knew how much Iwen and the other men distrusted and disliked the CIA. They believed the government was using the CIA in its secret plot for world control. Handing a list of agent names over to Taiwanese officials would put those agents out of circulation.

Mike phoned Burg with the news. "Iwen plans to get the list to me before I go to Madison," he said. "And I'm supposed to pass it on to Lewis."

Burg knew that passing such a list would be harmful to U.S. interests. Not only did it carry the potential for compromising the work of the CIA agents, it put their lives and the lives of their connections at great risk.

The FBI had to prevent the list from getting into the wrong hands. Fortunately, it was supposed to be passed through Mike first and that would give them a chance to intercept it safely.

On the Laser's Edge...128

The list gave the Friday meeting with Lewis even more importance. Depending on how it was handled, it could serve as another significant piece of evidence.

Agent Burg had already been spending most of his time that week putting the wheels in motion for the Friday meeting between Mike and Representative Lewis. Although he had received preliminary approval for the taping, he needed official authority as the day approached. He traveled to Milwaukee to seek it from Supervisor Byrne and SAC Hogan. As before, he would explain to them the investigation, the events leading up to the taping, and why it was necessary.

Byrne was now a believer in Agent Burg, if not in the laser case. He had seen the agent in action for several months. His research was solid and his credentials impeccable. Though he would never be able to grasp the H.G. Wells reality of lasers, he was willing to give the agent the latitude he needed to catch a corrupt official. He was quick to give his approval for the taping, and passed it on to SAC Hogan for his stamp of approval. Once given, Burg dictated memos of the approval and a teletype to be sent to FBI Headquarters. Because the taping would involve a high-profile subject — a sitting legislator — Headquarters had to be prepared for whatever flak might come their way. It would require their written authority to proceed.

Mike was so happy to have the case moving forward that he dared not think that the future taping was anything more than another positive step in the right direction. His confidence only wavered when Burg informed him that he would not be available that Friday to help with the preparation in Madison. "I have a conflict, but I'll set everything up and I'll go over the process with you before you go," the agent assured him, "I've asked the Senior Agent to meet you in Madison, Henry Curran. He's familiar with the case. Our technical coordinator from Milwaukee will get you wired up, Rick Prokop. They're both nice guys. They'll take care of you."

Mike didn't let on that he was disappointed. He had been parlayed from one agent to another and only Burg had given him a genuine sense of hope. Now, at another critical time, he was being handed over to someone else. His only comfort was in knowing that Burg's absence was only temporary. He trusted Burg, and in return, Agent Burg trusted him; at least, as much as an agent wants to trust an informant. A relationship based on colossal secrets is edgy at best. Too

Code Word Tikal...129

many of them collapse from the weight of suspicion before the long, laborious steps of collecting evidence concludes. To Mike's credit, he had a long and trustworthy track record, and he was no quitter. Agent Burg had a feeling he could place his bet in his corner.

The agent's schedule had become a flurry of activity. He had numerous other cases that needed his attention, and they all had to be shoved aside as much as possible so he could concentrate on the laser case. Immediately after receiving the green light in Milwaukee, he drove to Madison to confer with Tuerkheimer. They discussed the possible need for Grand Jury subpoenas, depending on the results of Mike's meeting with Lewis and the passing of the CIA names; then, it was off to the Madison office to discuss the Friday taping with Senior Agent Curran. Because of his expected absence, Burg needed to fill Curran in as best he could about what was going on and what they hoped to achieve with Friday's taping. Burg's last responsibility was to drive back to Wausau and prepare Mike for the task ahead.

Unfortunately, Mike had bad news waiting for Burg. It was about the CIA list. "We didn't connect," Mike told him. "Iwen never got the list to me."

Burg was disappointed, but realized they were simply back to where they were before knowing about the list. There was no reason to cancel now. He could only hope that enough ammunition would be provided in the tape to serve subpoenas, and to keep Mike out of Guatemala. Perhaps the list would surface later as an added bonus.

"Try to find out about the list," Burg prompted. "Where it's coming from and what they intend to do with it."

Burg also briefed Mike on the details of the wiring device and coached him in how to conduct himself during his visit with Representative Lewis. "Good tapes are made, not born," he believed and he told Mike, "You're there to ask questions. Don't do a lot of talking yourself." He told him to avoid noisy areas, preferring him to stay in the legislator's office. "If there's a radio on, or music in the background, try to get it turned off," he said. "The less background noise, the better the taping quality will be. And be sure to mention names. When you speak to someone, address them by name so they can be identified on the tape. There's a lot of body language involved with conversations and we won't have the advantage of that when reading the transcript."

"Is there anything in particular I should ask him?" Mike inquired.

On the Laser's Edge...130

"We want to know why he's working with Iwen, what he hopes to accomplish and what he thinks his role is in all of this. Ask him about his interest in lasers and the trip to Guatemala. What's in it for him?"

"I'll try my best, Tom. Thanks for your help. I really appreciate everything you're doing."

"I should be thanking you, Mike. You're doing a terrific job."

The next morning, Friday, October 6, 1978, Mike kissed his wife and daughter goodbye. "I'm just dropping off a couple of radiation detectors at the university," he told Pat, "but I have a couple of other stops to make. I'm helping the FBI with something, so I might be home a little late tonight." He wanted to tell her about the body wire, as he had wanted to tell her about so many other things that had happened to him over the course of the last two-and-a-half years, but he knew he couldn't. Secretly taping the conversations of a state legislator was no small matter. Knowing what was about to happen would only worry her, and knowing too much might put her in harm's way.

He climbed into his rusted white car. As usual, a St. Christopher medal was along for the ride. He said his prayer for safe travel, and pointed his Pinto towards Madison.

That Friday was a typical October day with overcast skies and a nip in the air. The colors of autumn had faded in the north, but progressively brightened as he made his way south. In many ways, the season was like his own life: beautiful in some spots, fading away in others. On the perfect side, he had a wife he loved dearly, a joyful young daughter whose innocence made him laugh, and a career that impassioned him. On the dark side were the forces of evil trying to suck out his spirit.

The miles seemed to go by in a blur with Mike's thoughts trying to focus on work, yet tugging him towards the unknown that lay ahead. He was doing everything in his power to do the right thing, yet there had been so many disappointments and so many temptations. Several months had passed since the U.S. attorney had entered the picture, and still no charges had been filed against anyone. He wondered if this day would bring his ordeal any closer to an end.

He arrived in downtown Madison with time to spare for his early afternoon meeting with Lewis. He found a parking spot in a public lot a short walking distance from the capitol building, and started down State Street. He was headed first to a building a block away from the capitol where he would be wired with the recording device. As

Code Word Tikal...131

instructed, he entered the building through a side door, then proceeded up a narrow flight of stairs to a private office. There, Mike was greeted by a thin, older man in glasses standing about five-feet-six.

"Mike?" the suited stranger inquired as he extended his hand.

"Yes," Mike responded.

"I'm Henry Curran, Senior Resident Agent, Madison," he said. "How are you today?"

"Fine. Just fine."

"Good. Did Tom fill you in about what to expect?"

"Yes, he did. He said there'd be someone else here to get me suited up." Mike looked around the room and saw no evidence of the recording equipment or the second man.

"Rick Prokop. He hasn't arrived yet. He should be here soon."

Mike was anxious to get to his meeting with Lewis and nervously kept checking his watch.

"Shouldn't he be here by now?" he asked.

"He's running a little late. He's driving in from Milwaukee. Don't worry, we still have time."

But Mike did worry. *What kind of sloppy work was this?* he thought to himself as the time continued to pass. *You don't come late to something this important.*

Finally, with just minutes to spare, Prokop arrived.

"Sorry," he apologized. "I got held up in traffic."

"You sure know how to cut it close," Mike made sure to point out.

"It won't take me long. Just a few minutes to get you wired. You'll need to take your shirt off."

From a brushed aluminum briefcase Prokop took out a hand-sized, Swiss manufactured Nagra stereo recorder. It rested in a sleeve that he strapped around Mike's back, between the shoulder blades. Two wires extended from the recorder to small, round microphones at each end. These were taped to either side of Mike's chest. Normally, Mike would have had access to an on and off switch, but Agent Burg never liked to take chances. "Things will go wrong if you let them, so control them," he believed. He had asked Prokop not to attach the auxiliary power switch and to tape the recorder button into the on position, out of Mike's reach. That way, Mike couldn't accidentally turn the machine off at the wrong time.

"All right, Mike, you're ready to go," he said about five minutes later. "You have three hours of tape. That should give you more than

On the Laser's Edge...132

enough recording time."

Mike quickly redressed in his shirt, tie and suit coat. As he turned to leave, Curran assured him, "We'll be here when you return. Good luck."

Mike's chest was pounding as he walked briskly back down State Street headed for the State Capitol, his heart still beating wildly from the last minute preparation. He was a man of precision and the delay had thrown him off balance.

The Capitol is a grand, historic building, designed like a hub with four spokes. The hub is an imposing granite dome. The spokes are four, two-story hallways filled with offices and meeting rooms. Construction began in 1906 and was finished in 1917. The final product possessed all the embellishments incumbent of a state capitol building of its day: hand painted murals, polished stone, and bronze statuary. Like the building, this taping would be historic: it would be the first time the FBI ever covertly recorded the conversation of a sitting legislator inside the state capitol.

Representative Lewis' office was on the second floor of the North Wing, room 334-N. Mike hurried up the steps, around the open rotunda, and down a corridor to a door marked with Lewis' name. Inside were two offices. He stopped at the reception desk in the front office. Lewis' regular secretary did not work on Fridays. Batzler and an unidentified man, who seemed to be working as an aide for Lewis, greeted him instead.

"Hi," Mike said. "What time do you have, Debbie?" He looked at the clock on the wall and noticed that he was actually on time. "Thought I was late," he said. "Later than I am."

He pointed to the clock. "Is that right?"

"Yes, I think so," Batzler said.

"How are you?"

"Oh fine, how are you?"

"Okay. A little exhausted running all those steps. Hard to find parking."

Representative Lewis appeared from his office. Dispensing with formalities, Mike jumped into the conversation. "Jim, got something for you." He returned to the legislator the copy of the letter from Fuentes he had been given at the September 22 meeting in Stevens Point.

"Something, huh?" Lewis responded. He reached out and took the letter, then turned back into his office, with Mike following.

The office was small but comfortable with the walls decorated with religious and motivational quotations: "VIRTUE" read one, and another, "Lord, what wilt thou have me do?"[51]

Lewis threw the letter into a wastebasket and assured Mike that it would be safely shredded. He asked the whereabouts of the second sheet of paper, the list of notes. Discovering that it was still in Mike's possession he told him to "Just throw it out, " because, "I don't want you to be involved that implicitly with me. I don't want anyone to come up here and say, 'Why do you have that?'"

It wasn't long before Lewis inquired about the CIA list. "Now, [Al] said that you had a list," he began.

"A list of what?" Mike asked.

"Names."

"Oh. He told me.... he said there were some CIA names you were going to have."

"*You* have," Lewis corrected.

"He didn't give them to me. He said he was gonna' mail them to ya'."

Mike explained that Al was leaving on his trip to Colorado and the two wouldn't be able to connect for several more days. "He's going out to see Colonel Roberts," Mike said.

"Has he left?" Lewis wanted to know.

"He's leaving tomorrow morning."

"Asshole. I was gonna' make that trip with him."

"Well, you probably should tell him, then.... You mean, to see Roberts?"

"Huh hum," the legislator confirmed.

Lewis was irritated that Iwen had let a traffic violation interfere with delivery of the CIA list the day before. "He sure screws up his life," Lewis said. "You know, he Mickey Mouses around with all this little shit and we miss...the big picture.."

The men began talking almost immediately about the October 10 party in Chicago. Lewis provided a treasure trove of details.

"I'm going on Tuesday.... the Chinese Government is celebrating its 67th anniversary of the foundation of the Republic of

---

51 "Rep. Lewis regrets 'mistake' in laser plot" by Arthur L. Srb, Associated Press, *Wisconsin State Journal*, August 29, 1979.

China...Taiwan," he began. "And I'm going to a reception in honor."

"Where?" Mike interrupted.

"In Chicago. In the Chinese Consulate and they're gonna' be puttin' me up overnight and this will be a friends-to-free-China type thing."

Later he would confide, "...They're putting me up for the night at the Holiday Inn down there. They don't have to. They asked if I would... be able to stay over night and have dinner with them. The reception should be 6:00 to 8:00, and then they're having dinner at 9:00 and they said they would be happy to see that [our] accommodations are taken care of."

From his desk, Lewis withdrew a white card and handed it to Mike. It was the official invitation to the Chicago events. Mike successfully concealed his elation. He memorized every little detail... dates, times, places... knowing the FBI would find the information useful. He returned the card and switched his brain seamlessly back into the conversation.

Lewis and Iwen had often discussed what countries, other than Guatemala, they might rely on for laser and laetrile sales, and where they might find refuge if needing to flee the U.S. Lewis was touting Taiwan.

"Now listen. It's important," Lewis said. "That's another country we can go to."

"Who?" Mike asked.

"Dealing with the top in Taiwan."

"We're goin' to Formosa?"

"Yah."

"Why? You know the top?"

"Yah, and that's why I want to keep all of these political contacts open, Mike, so that we have a sanctuary, wherever it might be, in case Latin America fails..." He said he trusted the Chinese Consulate, professing, "You see, here's the circle of friends I'm trying to move in, the World Anti-Communist League."

That he should choose Taiwan was no fluke. Lewis' committee work in economic development had allowed him to play an important role in establishing trade relationships between Wisconsin and the Asian country. Taiwan was also a rich source for the required raw materials needed in the making of laetrile.

The men went on to discuss Iwen's idea of using a relief shipment as their cover for their next trip to Latin America. Lewis was concerned about using a church minister who was unfamiliar with their underlying motives. "...that's fine, but the problem is, is what if [Al] can't get the pastor to go on to Salvador; if the pastor wants to hang around, you know? We're talking systems and not compassion... It's fraught with a lot of potential for...not giving us the safety valve that we need. ...It gives us an excellent... reason for going down there... a sister city program, the neighbor to neighbor concept, and all the rest of that shit, but Al has not been able to fine-tune a thing."

David P. had been suggested for filling the ministerial role. "...If we're gonna' go under a relief program and [David P.] comes in wearing his frock, I could give a shit. I could care less. That's beautiful," Lewis said, adding, "I've got a right to go down there as a legislator and I've got an even better right to go down there with these men of the cloth. That's swell."

It was not the only time Lewis would talk about using his political clout to help with the laser plans. He continued to emphasize his desire to travel with Mike on their upcoming trip as a way to assure the scientist's safety.

"For the next two years I can be your security blanket," he told Mike, referring to his term of office.

"What do you get out of it?" Mike interjected. It was one of the questions he had been coached by Agent Burg to ask.

"Well, I'm compelled by other motivations," the assemblyman started... "I want to do what's right."

Mike knew there was more to the answer. He had heard Lewis in other conversations talk about money and power. He would not let the topic die.

"Let me ask you a question," Mike said, directing the conversation down a different path for the time being.

"Yeah."

"What happens if they keep me there and then they get the technology they want and then you guys are ripped off? Did you ever think of that?"

"Yeah, I thought of that."

"What happens to that?"

"They won't keep you there because of me..." Lewis said, latching again onto the importance of his political position.

"No, I mean..."

"I want to go with you..."

"No, I mean, they would keep me there for the technology. That's all they want is the technology. But once they got the technology, once they have me, and they decide you're not useful, then what happens to you guys?"

"I am very useful to them for those two years... As long as I'm in public office, you're safe."

"How do you see that?"

"Well, as...an elected representative to a state government...I have a great deal of authority and clout."

"Okay."

"And they wouldn't want to upset that delicate balance between the Department of State, the Congress and the State Legislature, so it is my suggestion to Al that you and I travel together..."

Perhaps a better view of how Lewis perceived his role came later on. "All I am is a politician," he said. "I just know how to get things done. I'm a doer. I'm a leg man... I ran advance for President Ford in three cities."

"You did?" Mike asked. He was surprised at the information.

"You bet. I'd walk into a town. I wouldn't know anybody. There'd be five advance people and I'd be one of them. I was a member of the White House staff of President Ford."

"You were?"

"You bet."

"I didn't know that."

"Damn right. They want a doer, I'm a doer. There were 124 of us in the whole country. People that get things done... I can work with mayors, with governors and with legislators, and that's all I do. I'm an advance man."

Mike was happy with how the conversation was going, but he knew he still needed Lewis to admit what he expected to gain by his involvement.

"Okay, what's in it for me is the bucks," Mike started. "...what's in it for you?"

"Nothing," Lewis replied flatly.

"I don't...that's what I don't understand," Mike pressed on.

"There's nothing in it for me," Lewis persisted.

"Once you're down there? You mentioned a standing army and all that [when] we were out at the hotel, remember? You see that as real?"

Mike was reminding Lewis of his statements at one of the meetings in Stevens Point when he said, "I'd have an army at my disposal, and drive a Mercedes. I'd have my own airplane, and a ranch."

"Maybe," Lewis replied.

He would finally admit, "I want the money, too. I want to see the dollars." He hadn't dismissed the ranch either, eventually admitting that when war came to the U.S. and he needed to escape the country, "They [can] put me out on a ranch some place."

There was more, though, to the legislator's motivations. Like Iwen, he believed strongly in what he was doing and believed it from both political and religious points of view. Those views he would also reveal to Mike. "I feel a very strong Christian commitment," he began.

"Why do you think it has to be done?" Mike said, referring to the landmarks targeted for destruction inside and outside the U.S.

"I'm a believer."

"You said that to me once before. You said that you felt that it was a drive, something was pushing you into history."

"Yeah," Lewis agreed. "I feel God, the Holy Spirit, is moving me in this area. I'm not to be the standard bearer of a flag or anything, but I just feel that the overall part of God's plan is to uproot the evil in the fields and that this is something he's allowing me to move into at this time. He could mean imprisonment, and I'm willing to pay that price."

"...you feel that the whole system here, you're sitting right on top of it now...you feel it's really gone?"

"Ah hmmm," Lewis confirmed. "I feel it's gone."

"Why don't you guys just get up and tell the people it's gone?"

"I did, I did. I asked and nobody wants to listen to me."

"They won't listen."

"Ah hmmm." Lewis proceeded to explain that there was an international move towards communism, clearly visible in Nicaragua and its struggles. "...We've overlooked the fact that the Cold War is still hot," he said. "And that what we're dealing with is international communism..." It was obvious that he believed the U.S. had already fallen victim to the struggle, echoing the opinion of fellow extremists who were convinced that the inner circle of the U.S. government had moved too far to the left. Of the communists he said, "...they don't have to fire a shot to take over our country. They've got it. They've

On the Laser's Edge...138

got the Department of State, they've got the CIA..."

Lewis said he understood the world power structure, "... and I understand...that...the Jews are in charge, and Protocols of the Elders of Zion and all the rest of that shit.... but, you know, I'm afraid that by the time we're done, if we don't square things off, we've met the enemy and he's us because of our inaction..."

Mike had asked the assemblyman why he didn't feel he could go through the system to make changes. "Don't you think that if you just went through channels of the power level you're at now, you could turn it around...?"

"Never," Lewis said. "Never."

Throughout the conversation, Mike would ask the legislator about one of the various other characters who had surfaced in conversations or meetings: likely sympathizers or players in the laser plot. As he had suspected, the details had remained a closely guarded secret and few others had been allowed into Iwen's inner circle. "...the problem is, anybody stateside would be a devil," Lewis believed.

Of Colonel Archibald Roberts he said: "...I appreciate what he's doing, but I have a negative feeling ... I'm sure he's trustworthy and he's as loyal in patriotism as anyone else, but my concern is his lack of willingness to be involved with politicians, which is what I am. I can't help that. Now, anybody that says that I'm no good to this cause because I'm a politician can kiss my ass."

Mike explained Iwen's reasons for seeing Roberts that weekend. "He said he was gonna' see Roberts in case somethin' went wrong and wanted Roberts' backup. That's why he's going out there this weekend."

Lewis laughed. "What's Roberts going to do?" he asked. "Write a letter to the editor?"

Lewis didn't know what to think about Claussen.

"Maybe Claussen's an agent," Mike offered.

Lewis dismissed the idea. "He's playing games," he countered.

"Do you think so?"

"Yes sir. He's playing games."

"...For what?" Mike asked.

"I don't know. He likes it. Maybe he likes to read Mickey Spilane novels."

He said he told Iwen to get rid of Claussen. "...I told Iwen, get rid of the excess baggage because, I like John and he's a sweet guy, but he's excess baggage..."

One person Lewis did like and trust was Colonel Fuentes. Lewis said of him: "I really like him. I trust him as my buddy..." He had maintained phone contact with the Guatemalan official and supported that country's battles against guerrilla groups. He believed leftist groups were instigating the uprisings. He had also attempted to make contact with Somoza, the dictator of Nicaragua, drawing up a letter "to let Somoza know he has friends." He felt if Couture and David P. went to Nicaragua as planned, "They can go down there as bringing them my greetings. I did that so that they'd have another reason for seeing him." He supported the President of Nicaragua against international communism, believing that the freedom fighters, the Sandanistas, were a left-wing communist group.

Mike broached the topic of money by complaining about Iwen's various plans to finance the laser scheme, including the forging of connections with the Mafia. Lewis said he didn't have any particular connections with the Mafia, but they did support him politically and he was confident he could get their financial support if it was necessary.

"I never pursued that," Lewis said.

"I mean, were you still interested in pursuing it?" Mike asked.

"Hmmm," Lewis affirmed. "If I'm told to."

"Who do you mean, by Al?"

"If Al would ever want me to pursue it."

Mike had done well in asking all the right questions and pumping the state legislator for all the information the FBI had requested. All that remained was firming up any details about the pending trip to Latin America just days away.

"What about the 13th, Jim?" Mike asked. "I mean, is it together, or what's the deal?"

"Do you and Al want to make the trip?"

"Al wants to do it."

"We'll do it if you can provide them with the show that they need," Lewis replied, alluding to Mike's laser film.

It was already clear that Lewis' reference to the show had nothing to do with the ancient ruins. Earlier in the conversation, Mike had specifically sought clarification, and the legislator had replied, "Ah, we're not gonna' go to the ruins..."

Lewis echoed his promise to travel with Mike, expecting his role as a state official to give them protection. "I want to make that trip. I want you to travel with me...because I'm a legislator...and if we get hassled by anybody I can raise a stink..."

The time was nearing Mike's 2:00 appointment and the men began making their way towards the office door. Then, Lewis did something unexpected. He reached up and patted Mike on the back, directly between the shoulder blades. The laser scientist's heart skipped a beat as he felt the hidden recorder press closer to his back, drawing the microphone wires taut. The legislator kept on talking, apparently unaware that he had just missed hitting the recorder by a fraction. Mike held back a huge sigh of relief. In a bit of irony Lewis' final words were, "Thanks, Mike."

As Mike left the State Capitol, he stepped lighter than he had in years. He felt the conversation with Lewis had gone well. Assuming that the tape recorder had done its job, certainly the legislator had said enough to incriminate himself. Better yet, it would once and for all convince wary FBI agents that the bits and pieces of conversation he had been feeding the Bureau all these years was not a figment of his own imagination. How befitting, he thought, that in one of the most venerated buildings in the state, where laws were created, one of its lawmakers had just admitted to breaking the law.

As promised, Curran and Prokop were still at the State Street office on Mike's return. "How did it go?" they wanted to know.

"I think I got some good stuff," he reported calmly and quietly.

Mike removed his coat and shirt so Prokop could take off the recorder. Prokop quickly tested the machine to assure that it had in fact recorded the conversation. "It sounds good," he said. "I'll get it back to Milwaukee right away to get it transcribed. Tom will get a copy of the results."

Mike returned to Wausau with high hopes that the tape would lead to charges, not only against Lewis, but Iwen as well, and hopefully others. He could feel a great weight being lifted from his shoulders.

The feeling, however, was premature.

CHAPTER SEVENTEEN

Man is the only kind of varmint sets his own trap, baits it, then steps in it.
– John Steinbeck, Sweet Thursday

During the very contentious presidential election year of 2004, the possible exposure of a CIA agent set off bells in Washington and grabbed headlines. There was a lot of speculation about whether Valerie Wilson, the wife of a former U.S. ambassador, was really an agent whose identity needed to be protected, or just a garden-variety CIA employee who had no special right or need for identity protection. It was a politically charged issue because it was being used on both political sides in a finger-pointing exercise about who was responsible for the exposure, why, and whether or not it was important.

Politics and the Wilson case aside, the point was made that it is illegal to expose the identity of an undercover agent. Doing so can be dangerous for the agent, and for anyone he or she is involved with.

Iwen claimed that he had the means to expose not just one agent but an entire list of them. What must the CIA have thought when they found out?

It took several days for the news to reach anyone in Washington D.C. Burg's report on the Lewis taping had fallen between the ubiquitous cracks of the system, placed there by the weekend and a government holiday — Columbus Day — on Monday.

On Tuesday, all hell broke loose. FBI Headquarters finally reviewed Burg's report. The list was supposed to be passed to Taiwanese officials at a meeting in Chicago that evening.

Burg received a call that morning at his office. On the line was a Bureau supervisor in Washington who wanted to know more about Representative Lewis and the CIA list. When he heard the details, the supervisor insisted, "You have to prevent this list from being passed. Get to Chicago."

Burg knew his place in the system and knew he couldn't take his orders from anyone but SAC Hogan.

"I understand," the supervisor said. "I'll handle it. Just get to Chicago."

Only minutes later Burg's phone would be ringing again, this time from Milwaukee. SAC Hogan was on the line. "What the hell is this all about, you going to Chicago? What's up with the CIA and Taiwan?"

"I sent a teletype on Friday with the details," Burg explained. "It had to pass through Ray first so I assumed the two of you would have discussed it."

"No," he had to admit. "So what's going on?"

"Representative Lewis was expecting a list of CIA names to come through the mail from one of his contacts up here. They were the names of some Taiwanese agents. He was planning to pass them off to the Chinese at some gathering this evening in Chicago. Headquarters wants me to help intercept."

"Well, then, I guess you'd better get going." Hogan replied. "I'll talk to you when you get back."

"Does Ray know what's up?" Burg asked.

"I'll deal with Byrne," Hogan said. "Just get to Chicago."

Burg rushed back to his home to pick up a clean shirt and tooth brush. His wife Pat had not left for her teaching job at North Central Technical Institute and was there to send him off. As a former field office secretary, she was accustomed to his unpredictable schedule.

"I don't know when I'll be back," he apologized. "Probably tomorrow."

"I know the routine," Pat assured him. "Good luck."

Minutes later he was heading to the airport to catch a flight to Chicago.

Less than an hour after takeoff from Central Wisconsin, Burg's plane touched down at O'Hare Airport. At the gate two agents from the downtown Chicago field office met him, sent there by security squad supervisor Ray Wickman. Headquarters had already alerted Wickman, a friend from Burg's days on the job in Illinois, about what was going on and what needed to be done. He had his agents bring Burg directly to his office so they could discuss it.

"Hi, Tom," Wickman greeted him. "Great to see you again. Looks like you've been feeding well on the Wisconsin cheese."

"Too well, I'm afraid. I'll have to shed a few pounds before the next physical."

"Since talking to you on the phone we've been able to pull some information together. This party tonight is at the Chiam Restaurant in Chinatown."

"I've been there. Nice restaurant."

"Lewis is registered at the Holiday Inn on the lake front north of the Loop."

"Fast work. So, what's the plan?"

"Well, we don't know Lewis' schedule, or where he might try to pass off the list, so we're going to post agents at both places: the restaurant and the hotel. Wherever Lewis shows up first, agents will be there to pick him up."

"If he doesn't have the list with him, we might have a problem detaining him."

"That's all right. We don't have to arrest him unless things get out of hand. We'll just ask him to come in for questioning. If that doesn't work, we'll have to pressure him. Legislators think they can talk their way out of anything, so I'm betting he'll come peacefully."

That evening, Burg and Wickman whiled away the hours at the Chicago office talking about the case and old times. Meanwhile, Chicago agents were patiently waiting for the arrival of Lewis. Not knowing where to expect him first, some were positioned outside the Chiam Restaurant, some inside at the reception, and others at the Lakeshore Holiday Inn. For added measure, the FBI reserved a room at the inn, next to the one reserved for Lewis.

At around 6:00 PM, several hours into the watch, Agents Jim Nealis and Frank King hit pay dirt. They spotted Representative Lewis' car circling the block outside the restaurant. The legislator was looking for a parking place, so the surveillance agents obliged, pulling one of their cars away and leaving a space directly in front of the restaurant. After parking, Lewis and his girlfriend stepped out of their vehicle and started towards the restaurant. The couple was dressed for the occasion: Lewis looking his usual dapper self in a fine suit and accessorized by Batzler in her eloquent evening dress.

According to plan, Nealis and King let the couple enter the reception. The inside agents would keep a watch on the activities, paying particular attention to any movement that might indicate that the CIA list was being passed.

Lewis worked the room, sipping drinks and talking. There was no indication that the event was anything more than a polite get-together

On the Laser's Edge...144

among dignitaries.

Then, at around 8:00 PM, Lewis showed signs of getting ready to leave. He and Batzler made their way towards the entrance, but not alone. The wives of three Taiwanese officials accompanied them. The agents, confident that the small group was headed for the 9:00 semi-private dinner with the consulate general, decided to move in.

"They're on their way out," one of the agents inside informed Nealis and King by radio.

"Well, this is our best chance to make a move," Nealis said. "Let's make it a good one."

Nealis and King got out of their car and hurried to the entrance in time to catch Lewis, Batzler and the three Taiwanese women as soon as they appeared outside the front door. Following close on their footsteps were the agents from inside the reception. The four agents cut into the group and positioned themselves between Lewis and Batzler. The Chinese women screamed, startled by the sudden appearance of the strangers.

"Representative Lewis?" Agent Nealis said as he locked his hand onto the legislator's arm and pulled him aside.

"Yes?"

"We're with the FBI. Would you come with us?"

"What's wrong?"

"You're not under arrest," Lewis was assured, "We just have a few questions we'd like to ask you."

Without protest, Lewis was escorted to a waiting car, and Batzler to another, each accompanied by two agents.

Back at the Chicago Bureau, Burg and Wickman heard the crackling interruption of the two-way radio, then the voice of Agent Nealis. "We have the subjects detained," Nealis reported. "We're bringing them in."

"All right!" Agent Burg exclaimed in triumph. "Now it's show time."

The Chiam Restaurant was fifteen minutes from the downtown offices of the FBI. On arrival at the Dearborn Street building, Lewis and Batzler were escorted from a secure basement parking area, up a short flight of stairs and to an elevator. In the elevator, the couple was briefly reunited for the trip up, then again separated on the tenth floor. Agents Burg, Nealis and King interviewed Lewis.

"Representative Lewis, I'm Tom Burg, Special Agent FBI," Burg introduced himself. He specifically avoided any mention of his

Wisconsin ties in an attempt to keep Lewis off-balance and from connecting the dots back to Mike. "Before we ask you any questions, you should understand your rights." The agent proceeded to recite the Miranda rights, as required even in non-custody situations at the time. When finished he asked the legislator, "Do you understand these rights?"

"Yes."

"Then I must ask you to sign a Waiver of Rights. It is a copy of what I have just told you and confirms that we have advised you of your rights and that you are answering on your own free will."

Although Lewis wasn't in custody, the tone in Burg's voice left no doubt that this was not a casual questioning where answers were optional.

Agent Nealis cut to the chase.

"I believe you have a list of names you were about to provide to the Chinese," he said. "If you have it, we would like to take a look."[52]

Lewis reached into his coat pocket and pulled out a three-by-five inch index card. He handed it to Nealis. On it were seven Asian names. They meant nothing to any of the agents present, but they knew they would likely mean something to the CIA.

"Where did you get these names?" Burg asked.

Lewis said they came from Albert Iwen, President of United States Pharmaceutical Incorporated (USPI). "The letter was on USPI letterhead addressed to me and stated, to the effect, that 'This is a list of CIA agents' and a list of seven names followed," Lewis explained. He had copied the names onto the card and placed the original letter into his office safe at the capitol.

"Why did he send you the list?" the agents wanted to know.

"I'm not sure. I called Iwen on the afternoon of October 9 because I was returning his call. He asked me if I had received the letter. I responded that I did and asked him 'What was I suppose to do with it?' He responded that he thought that I would be interested in the information."[53]

"So, why did you bring it to Chicago with you?"

---

[52] From court exhibits A & B, case #79-CR-57, *United States versus James R. Lewis*, U.S. District Court, Western District of Wisconsin.

[53] From court exhibit "B", statement to the Federal Bureau of Investigation, case #79-CR-57, *United States versus James R. Lewis*, U.S. District Court, Western District of Wisconsin.

On the Laser's Edge...146

"I had an intuition that the FBI would ask me about it."

"An intuition," Burg repeated skeptically.

"Yes."

Burg knew that Lewis had just taken a deadly step. He not only admitted to knowing about the list, he was to trying to justify his possession of it by explaining that he had plans of turning it over to the FBI. It was the kind of excuse that seldom worked. What Lewis did not know was that the tape recording in Madison would prove him wrong. The questioning continued for more than two hours. Burg took the lead. What do you know about Albert Iwen? Do you know John Claussen? What about Colonel Archibald Roberts? The 1313 building in Chicago? The Posse Comitatus?

When asked about Iwen, the legislator called the chemist and laetrile manufacturer "a difficult man to get to know. Either he's crazy or he's a genius."[54] He said he sometimes felt uneasy around him, but continued to meet with him at least once a month because he supported laetrile and the work that Iwen was doing. He even supported Iwen's efforts to set up a laetrile plant in Latin America because he believed in the possible health benefits of the drug. Yes, he had taken a trip to Latin America in the spring with Iwen to explore manufacturing options there. And yes, he was planning another trip to the area on the upcoming weekend.

Except for the CIA list, Lewis seemed forthright. But, as the evening progressed, the questioning intensified. Like a fisherman who had already set his bait, Burg began playing his line. What do you know about lasers? Did you ever discuss terrorist attacks on Cuba or O'Hare Airport? Who is Mike Muckerheide? Wasn't Tikal a reference to a laser plot?

By this time, Lewis must have realized the awful truth: the FBI already knew too much. But how much they knew, and how they had come to know it, was still a mystery.

In another room, Batzler was receiving a similar interrogation. As expected, her knowledge about the laser plan seemed limited. She admitted that she didn't particularly like Iwen and often found his conversations disjointed and confusing.

At around midnight, the questions ended and the Chicago agents drove the couple back to Wisconsin: Batzler in Lewis' car with two

[54] FBI interview notes dated 10/12/78, MI 2-30.

agents, and Lewis in a Bureau car, also with two agents. The six arrived at around 3:00 AM at the Wisconsin State Capitol in Madison. The state representative registered in the capitol sign-in book, then led the party to his office upstairs. At the request of the agents, he opened his safe and handed over the original letter from Iwen that included the list of CIA names. Dated October 6, 1978, the letter was addressed in capital letters to THE HONORABLE JAMES R. LEWIS and read:

> *DEAR JIM,*
> *The list of known CIA agents in Nationalist China is as follows:*
> *[names deleted]*
>
> *(All thought to be Army Generals)*
> *American "employees of Air Asia"*
>
> *Affiliates of: American Express and International Banking Corp.*
> *Military branch.*
> *Sorry I didn't get enough time to send this list along with your visitor for this Friday.*
> *As ever,*
> *Al*
> *P.S. I'm certain this list is incomplete and is about eight years old.*

"Thank you for your cooperation, Mr. Lewis," Nealis said. "But we still need your passport." Lewis had already been told not to expect any more trips out of the country for awhile.

"It's at my home in West Bend," Lewis said.

"Then you'll have to go get it. You are ordered to appear personally at the FBI office here in Madison to surrender the passport by the end of this business day."

The four agents left at around 3:30 AM. Both Lewis and Batzler were left with no chance of getting a good night's sleep, and with subpoenas to appear before a Federal Grand Jury in Madison.

CHAPTER EIGHTEEN

If anything ail a man, so that he does not perform his functions, if he have a pain in his bowels even...he forthwith sets about...reforming the world.
— *Henry David Thoreau*

Life for Mike had to go on like the secret taping at the State Capitol had never occurred. He couldn't even allow himself to see the glimmer of hope it promised for the end to the long dark tunnel that had become his life. He had to continue pretending that nothing had happened.

Fortunately, he had the weekend to mentally recoup. Iwen had flown off on his rescheduled Colorado trip on Saturday, and Mike would not be in contact with him again until Monday. Even then, it was only a brief conversation by phone. Iwen reported that he had successfully made the trip to see Colonel Roberts, it had gone all right, and he wanted to know if Mike was still able to make the trip to Guatemala. After Mike said yes, Iwen placed a call to Fuentes, likely to confirm their upcoming visit.

The next day, October 10, Iwen went to a travel agent in Wausau and handed over $1,077 in cash for round trip air reservations from Chicago to Guatemala City for three individuals: himself, Lewis, and Mike. He asked the travel agent not to mention the trip to anyone.

The trip was just three days away and Mike was getting concerned. He had talked to Agent Burg on Friday night to report what had happened in Madison. When he called Burg on Tuesday to tell him about Iwen's ticket purchase, Burg was nowhere to be found. Four days had passed since Mike's taped conversation with Lewis and nothing seemed to be happening towards bringing the legislator to justice. Knowing that it had been the Columbus Day weekend was of no consolation. Crime did not take holidays, and by Wednesday, Mike was wondering again about his fate.

Unknown to him, that morning, Agent Burg was still in Chicago, basking in success. He had stayed behind to finish up the paperwork and to prepare for a busy day ahead. It was Agent Wickman who contacted the Bureau Supervisor in Washington to let him know they had retrieved the list of seven CIA names. The supervisor was elated and congratulated Wickman and the other agents: "We gave you an impossible assignment," he said, "and you did it."

At around 3:00 AM, Burg had finally retired to his hotel room at the Palmer House a block from the Chicago field office. He would only catch a couple of hours of sleep before rising early to check in with Wickman and catch an Amtrak train to Milwaukee.

Jubilation for the successful mission continued at the Milwaukee office, but the celebration was short-lived. There was still a lot of work to be done, and the first was assuring the delivery of additional Grand Jury subpoenas. Agents delivered nearly a dozen over the course of the next two days from Wisconsin to Colorado, most notably to Iwen, Claussen, and even Mike Muckerheide. "We have to subpoena you too, Mike, for your own protection," Agent Burg had explained to him. "It will give the other guys the idea that you're still one of them, and your testimony will give the Grand Jury an honest view of what these guys were trying to do."

Interviews coincided with the subpoena deliveries. Over the course of the next twenty-four hours additional agents from Milwaukee were sent to Wausau to help with the task. The goal was to conduct interviews quickly and simultaneously so there would be no time to compare notes and concoct fictitious excuses.

Milwaukee agents were assigned to conduct interviews with Iwen and Mike, while Agents Burg and Szekely were assigned to Claussen. The reasons were simple: if the Milwaukee agents interviewed Iwen it would send a message that he had again caught the attention of someone outside local law enforcement. He would likely take the interview more seriously.

Misery loves company, so having that same set of agents visit Mike would reinforce a sense that he was a fellow victim. The FBI still needed his cooperation, so giving him preferential treatment would have aroused suspicion.

For the opposite affect, different agents were assigned to Claussen to isolate him from his comrades. That would hopefully raise suspicions and start dissent. What's more, they felt Claussen might cooperate. Since Burg and Szekely knew the case better than anyone, they stood a better chance of steering the financial planner into their camp and being there to follow-up.

The interview with Iwen was first on the agenda. Aside from Representative Lewis, he was the key figure — the linchpin — and the agents were anxious to hear his version of the story.

The agents found Iwen at home in Merrill, living with his elderly parents in a small, older frame house on Fifth Street. If they expected a lot of information, they were disappointed. Iwen wouldn't budge. If his troubles with the law had taught him anything, it was the art of passive resistance.

As the agents fired their questions, Iwen remained tight lipped. Without glorification he answered only a few questions. He said he had little knowledge about laser weapons, though he knew they existed, and didn't know anyone connected with the military in Latin America. He did meet someone named Fred Fuentes during a trip to Guatemala, but only as a result of Lewis' interest in photography.

The information wasn't a lot, but it was a start, and the agents knew it locked Iwen into a position that might be later exploited.

If the FBI hoped to gain anything more, it was going to have to come from other sources. Claussen was their next, best hope.

Burg and Szekely arrived at Claussen's country home east of Wausau the following afternoon. Claussen seemed to be doing all right financially, the agents noted as they drove up to his residence. He had a newer, though modest, ranch style home in a nice, quiet neighborhood. The two agents knocked at the door.

"John Claussen?" Burg inquired.

"Yes."

"I'm Tom Burg and this is Richard Szekely. We're special agents with the Federal Bureau of Investigation. We'd like to talk to you. May we come in?"

Their visit would not have come as a surprise to Claussen. Word had already spread about the fiasco in Chicago and Iwen's interview in Merrill. But, the men would not have been able to connect all the dots yet. They would have been left wondering what the FBI was actually investigating — lasers, laetrile or Lewis?

Claussen stepped aside and allowed the two men to enter. He led them down a hallway and into a neatly furnished living room. The agents took a seat and advised Claussen of his rights. Claussen didn't ask to have his attorney present, but he did want to know: "Are you taping this interview?"

"No, we're not." Agent Burg replied. "Are you taping the interview?" It was Burg's attempt at controlling the conversation: the two agents wanted the upper hand; they wanted to be the interrogators, not the interrogated.

"No," Claussen said and the agents continued.

Easing into the conversation, the agents asked simple, personal questions at first: date of birth, education, employment background; then, drifted into deeper territory: what is your association with Albert C. Iwen? James R. Lewis? Myron C. Muckerheide?

Claussen admitted to knowing Iwen since 1972 when they were both involved in developing the American Party locally. They still occasionally had coffee together, and, yes, he knew about Iwen's interest in laetrile, but had no business association with the chemist.

Likewise, he said he knew James R. Lewis. Iwen had introduced him for the first time that spring in Stevens Point. Over lunch, they had talked politics and laetrile, nothing more.

As for Mike Muckerheide, Claussen seemed particularly evasive. He admitted to knowing the laser scientist, but didn't know much about what he did other than what he read in the local newspapers. When asked why he was being evasive, Claussen said he didn't want to get Mike into "a mess of things."

The topic turned to guns.

"Do you have any guns?" one of the agents asked.

As with many of their questions, Claussen followed with a question. "Is there anything wrong with having guns?"

This time the agents didn't press the issue. Claussen's wife had entered the room with a young child. Sometimes a spouse was a hindrance and sometimes a help to an investigation and they weren't sure which Mrs. Claussen might turn out to be. They didn't need to take chances, either, by angering her with a heavy-handed interrogation of her husband in front of the child. It wasn't necessary, anyway, since Claussen still faced a Federal Grand Jury inquiry. By soft-pedaling their own interview, they might be able to lead Claussen over to their side.

As the conversation continued, the agents noticed the changing expressions on the face of Claussen's wife. It had started as dismay, turned to alarm, and then to anger. Occasionally, she injected comments or questions of her own, directed at her husband. It was obvious that she knew nothing about her husband's extracurricular activities and was not pleased by the insinuations. The agents were secretly delighted.

"Are you an agent or an informant for another agency?" they finally asked.

Claussen refused to answer.

"If you are an agent for any U.S. agency, you'd better tell us, or tell the agency to contact us as soon as possible. This is a serious situation," Burg informed him. He handed him a business card and told him, "We'd like your cooperation. Call us if you have anything you would like to tell us."

Agent Szekely served Claussen a Federal Grand Jury Subpoena, explained his mandatory appearance in Madison October 25, and the two agents thanked him for his time and left. They drove off in their Bureau car smiling. Now that they were gone, they knew he was facing the toughest interrogation of all, at the hands of his own wife. "He no longer has the right to remain silent," Agent Szekely chuckled, as they headed back down the road towards Wausau.[55]

---

[55] Details about the questioning of John Claussen taken from interviews with Tom Burg.

CHAPTER NINETEEN

*"...allow me to say that, if I cannot get an opportunity
to try my new mortars on a real field of battle,
I shall say good-bye to the members of the Gun Club..."*
— *Jules Verne, From Earth to Moon*

Trust is a fragile thing, and the trust the four men (Iwen, Lewis, Claussen and Muckerheide) had slowly built, was beginning to crumble. In the coming days before their Federal Grand Jury testimony, they stayed in contact — comparing notes, and wondering about their fate — but not without each of them harboring an uncomfortable suspicion towards the others. Mike never doubted that his cover would one day be blown: it was just a matter of controlling when it happened. Now was not a good time.

His first face-to-face contact with Iwen since the Chicago incident was over a week later. The two had agreed to meet at one of their usual meeting sites, Sambo's Restaurant. When Iwen didn't show, Mike drove off to a nearby gas station to call the FBI office. On his way, he spotted Iwen driving towards the restaurant. He proceeded to make his call. It was Agent Szekely who answered. "Can I speak to Tom?" Mike asked. In the background was the familiar sound of symphonic music. Mike could imagine Zeke sitting at his desk puffing away on his pipe, dressed in one of his favorite sweaters, and on his feet, a pair of hiking boots.

"Sorry, Mike," the agent said between puffs. "He's not available."

"Well, Tom knows I was supposed to meet Iwen this morning at 7:45, but Iwen didn't show. I waited almost an hour. Then, as I was on my way over here, I saw his car. He's headed towards Sambo's right now. If you can, let Tom know I'm on my way back to the restaurant."

"Okay, I'll let him know, Mike."

Driving into the restaurant parking lot, Mike spotted Iwen's Mercedes parked in back. He pulled his Pinto into a nearby parking

On the Laser's Edge...154

spot and went inside the crowded restaurant.

"Thanks for coming," Iwen said.

"Why are you so late?" Mike countered bluntly. "I was here earlier and left."

"I'm just a little late."

"Well, what's up?"

"My lawyer wants you and John to take polygraph tests."

"Polygraph tests? What for?"

"The FBI knew about the head of Reich thing. They asked me, 'What's this about being the Fuerher?' They know things they couldn't possibly know without some inside help."

Mike decided to play hardball. If Iwen was going to accuse him of being the snitch, he would turn the tables and try to divert his attention.

"Listen, Al, when I met you, on the first meeting, you offered me a job. You told me, if I had some force field technology, you knew of a foreign country that was interested. This is all your idea, remember? I don't know what's going on here, but I never expected it to come to this and I'm not too happy about it."

"I don't think it was you, Mike," Iwen assured him. "I think it was Claussen. He was over at my house on Saturday. He flipped his wallet open like a Fee Bee and said the Fee Bees were such nice guys. I told him to stop fooling around, but I think he really believes it. He thinks they're nice guys. Did you talk very long to the FBI the other day?"

Mike shook his head. In fact, he had not. His lengthy cooperation with the FBI had made the interview unnecessary. Their visit to his house had been for appearances only.

"I didn't either, but Claussen said he talked to them a long time."

Mike heard Iwen's voice turning cold.

Iwen's concerns were not entirely misplaced. After his interrogation by Burg and Szekely, Claussen had started to question what he should do. That same day he had picked up Agent Burg's business card and dialed the phone number listed. He didn't ask directly what might happen to him if he decided to cooperate, but the agents knew he was testing the waters.

"I often wondered about Claussen," Mike said, encouraging Iwen's distrust. "I thought he might be CIA. I even asked him about it once."

"He didn't show any fear when I approached him about the leaks, but he did blush," Iwen said.

"I didn't blush when you confronted me, Al," Mike said, thinking back to the incident in Radtke Park. "I turned white." He wasn't lying.

Mike had been glancing out the window when he spotted Agent Burg's car passing the restaurant on the street out front. It was a welcome sight. Maybe it was again just his paranoia, but he felt there was something stranger than normal in Iwen's demeanor that day: a harder edge.

"So, what about the Grand Jury subpoena?" Mike plowed ahead. "What's going to happen?"

"I talked to my attorney. We're pleading the Fifth. He suggested that you and I stick together, use the same lawyer. He'll represent both of us."

It wasn't the first time that sharing attorneys had been proposed. Claussen had also talked to Mike about it, hoping to share the burden of legal fees.

"I don't know, Al. I don't know that that's such a good idea. I have a lot to protect: my career, my family. I've never been involved with something like this."

"I have another Grand Jury date coming up on the laetrile. My attorney knows how things work. He knows how to work the system. It would be good if he represented us both."

"I just wonder what will happen. What could happen?"

"Well, whatever happens, don't try to get a military contract after this because the FBI will be all over you. You have a record with them now, no matter what happens."

Mike chuckled to himself. He had quietly been doing some military work for the U.S. government for several years. And yes, he had a record with the FBI: a very long one.

"What about Lewis?" he asked. "Have you talked to Jim lately?"

"Yeah. He's really mad. When they were interrogating Debbie down in Chicago, the Fee Bees told her what a nice guy she had, that he was always shacking up with other women. I think Debbie and Joan will both get immunity. And even though we might not be able to do anything with the lasers, it doesn't mean we have to give up our plans in South America. We can use sophisticated electronics. It's inadmissible as evidence and they'd just throw it out of court."

"Are you going to be seeing Claussen again?"

"I'm supposed to see him tonight. But, he can come to me."

"I don't want to see him again. I've had it with John."

On the Laser's Edge...156

Mike had wasted much of his morning away on Iwen and he was anxious to get back to the comfort zone of his lab. "Listen, I have to go," he said finally. "I have to get to the Institute. Take care of yourself, Al."

When Mike left Sambo's, he saw that Agent Burg's car was now parked on a side street.[56] Aware to keep his distance, he left and reported to the agent by phone from his laboratory.

Staying in touch with the agent had taken on a new importance for the case and for Mike's sanity. With the Grand Jury testimony just days away, past, present and future were converging on a collision course, with Mike in the center. Even with the FBI protecting him, his life experiences told him that things didn't always go as planned. He trusted Agent Burg, but he didn't trust the bureaucracy to always get things right.

Thankfully, Mike got a brief rest from the case when the central element of his problems left for a long weekend: Iwen headed for Guatemala. With everything that had happened in recent days, Mike had a good excuse for not risking the trip, as did Representative Lewis whose seized passport had not been returned. Iwen, however, was not about to let the threat of a Grand Jury probe totally ruin everything he had worked so hard to create. Colonel Fuentes had recently told him that he now had the support of his government for laser weapons, with a $300,000 financial commitment. If things went wrong, the Guatemalan had promised to burn all the incriminating documents. For Iwen, there was still hope of salvaging his plan, and the sooner he could get positioned out of the U.S, the better his chances. He left for Central America for the weekend, taking David P. and Couture with him.

If he was relieved to be left alone for a few days, Mike's spirits were further buoyed with news that his first patent had been accepted. His invention relating to a laser system and method for diagnosing a condition in an inaccessible area, filed in September 1976, was officially allowed on October 17, 1978 and assigned to A. Ward Ford Institute. For a moment, Mike's up and down world was back up again.

---

[56] Details of meeting between Mike Muckerheide and Albert Iwen taken from handwritten notes by Mike Muckerheide dated October 16, 1978.

CHAPTER TWENTY

*A fanatic is one who can't change his mind
and won't change the subject.*
— Winston Churchill

It was the oddest conversation Mike ever had. Iwen was starting to come unhinged and he was sure he could hear the screws dropping. It was 7:52 PM and he was at home when he got the call.

"I want to talk to you, Mike. Is it okay to talk to you on the phone?" Iwen asked. He sounded anxious.

"Sure, it's okay. What's up?"

"I found out some information. I…I don't know. Colonel Roberts, he…he received some paper too and, well, it named me."[57]

Iwen's sentences were disjointed, but so far, Mike was following the conversation. The chemist was concerned that Colonel Roberts had been subpoenaed. Some of the information that the colonel had received, however, specifically named Iwen in a way to suggest the FBI knew more about the laser plan than he realized. He was falling into a tailspin.

"It's mortal, Mike, the source is mortal," Iwen said.

Mike didn't know what he was talking about. "Mortal? What do you mean?"

"Human," Iwen explained, before rambling on. "John Claussen is a good boy, Mike. He's good. You'd be surprised at how I found out. I don't know. You'd be surprised at my source."

He told Mike he wanted to meet him at the Country Kitchen in twenty minutes; otherwise, in the morning at 8:00, at Sambo's.

Mike made his decision quickly. "I'll meet you in the morning," he said. He wanted to avoid any more after-dark encounters with Iwen, who seemed more peculiar than normal.

"You'd be surprised at how I found out," Iwen kept saying. "I

---

[57] Details of phone conversation between Mike and Albert Iwen taken from handwritten notes made by Mike Muckerheide dated October 23, 1978.

On the Laser's Edge...158

don't know. I can't say on the phone."

"Have a good night," he concluded. There was something ominous about his farewell.

"Yeah, you too."

"Have a good night," he repeated.

It was hard for Mike to have a good night after a conversation like that. Even if he didn't understand everything Iwen had told him, he did understand that he was acting like a desperate man. If he really believed that his laser plan would save the world from communism, then what was he thinking if he feared his plan was being destroyed?

The next morning, Mike was up and out early, first checking in with the FBI before leaving his house. He spoke to Agent Szekely, informing him about his morning appointment with Iwen. With the case headed for a Grand Jury the following day, it was important to have an agent posted nearby. Since Milwaukee agents had interviewed Iwen two weeks earlier, only Claussen could identify the two local agents and he was running scared, seldom meeting again with the group. Knowing that Claussen was not likely to show up, Szekely left his office, headed for the Country Kitchen.

By intention, Mike was early for his meeting. He had allowed time for Szekely to arrive first, restlessly cruising down Grand Avenue to kill time. When he finally entered the restaurant he noticed the agent was in place at the counter. Following instructions, he sat down within hearing range, taking a seat at a booth just behind and to the left of Szekely.[58]

Iwen came in wearing a plaid cap. "Morning," he said cheerlessly, removing his cap and sitting down. Already, Mike noticed the change. He had seen the many colors of Iwen's moods, but not this one: the color of dark desperation.

"Coffee this morning?" the waitress asked.

Iwen placed his order and waited impatiently for the waitress to leave.

"Why did you need to see me this morning?" Mike asked.

"I've been to a clairvoyant," he said at last. "I was told there was somebody turning us in."

Mike knew that Iwen had become obsessed with discovering the

---

[58] Details of the restaurant meeting between Mike and Albert Iwen taken from handwritten notes made by Mike Muckerheide dated October 24, 1978.

source of FBI leaks, and he knew that the chemist had used clairvoyants in the past, but he had not expected this.

"The clairvoyant says it's a young girl," Iwen continued. "You were right about women, Mike," alluding to the laser scientist's earlier comment about avoiding the three-Bs: booze, broads and bucks.

"So, who is it?" Mike asked.

"I think it's Debbie. You and John are okay."

Mike was expecting to hear his own name. It was cathartic to discover that Iwen still didn't know the truth; but, it gave him little comfort. Iwen was a man of science and Mike knew that the clairvoyant was just a desperate detour in his search for the truth.

They talked about the Grand Jury investigation scheduled for the following day. Iwen did not seem especially concerned. One could suppose that by this time he had been involved in so many lawsuits that he viewed them as irritating inconveniences.

Before the two men departed, Iwen worried again about the leaks. "It's Debbie," he repeated. "I'm sure it's her."

Mike could hardly believe his luck. Iwen was totally blind to his deception; yet, he knew that time was running out on his secret. He only hoped that when the truth was told, he would have some control over the time and its terms. The uncertainty of it left an empty pit in his stomach. Driving away from the restaurant that morning, Mike was sure of only one thing: he would never trust his future to a clairvoyant.

CHAPTER TWENTY-ONE

As soon as the jury had a little recovered from the shock of being upset, and their slates and pencils had been found and handed back to them, they set to work very diligently to write out a history of the accident, all except the Lizard, who seemed too much overcome to do anything but sit with its mouth open, gazing up into the roof of the court.
— Lewis Carroll, Alice in Wonderland

The circus had arrived in Madison at the Federal Office Building. In the hallway, outside the Grand Jury room, several witnesses sat on long, oak benches waiting for their turns to testify, while others paced impatiently in the hall. Gathered together was an unusual group of talented people who were individually aggressive, outspoken and flamboyant. Many of them made no secret of the fact that they did not believe in the system and had their own ideas of how to change it. Mike watched as John Couture "walked around like he was God," and when Colonel Roberts' arrived, his wife ran up and down the halls announcing his arrival, "Colonel Roberts is here! Colonel Roberts is here!"[59]

As Mike sat among the well-groomed flock of witnesses, he couldn't help but notice their similarities. Except for Batzler and Joan C., they were all middle-aged men who were high achievers from disciplined backgrounds. For whatever reasons, many of them had turned their backs on convention and were seeking a greater place in history.

The taking of their testimony would be different from a traditional court case. First, witness attorneys would not be allowed into the grand jury room. The room itself would be laid out like a conference room. Everyone would be sitting around a large table: the jury members, the court reporter, U.S. attorneys, and individual witnesses. The

---

[59] Interview with Mike Muckerheide.

witnesses would be brought in separately for their confidential testimony. On this day, the U.S. attorneys at the table were Frank Tuerkheimer and his assistant Judith Hawley.

As Mike waited impassively for the process to begin, various members of the testifying cast approached him in the hallway. Iwen's attorney, who noted that Mike hadn't brought legal representation, came first. For this occasion, Iwen had hired Joe Wiegel, a small, thin man noted for his clientele of Posse Comitatus members. "Don't worry," he assured Mike. "We'll take care of this."

Next came Representative Lewis who motioned him off to the side. "We need to stick together," he said, lowering his voice. "If we all say the same thing, what can they do?"

"What do you want me to tell them?" Mike asked.

"As far as you're concerned, we never talked about lasers. In Chicago, I told them I didn't know you were a laser scientist. I told them I thought you were an urban planner."

"You told them I was an urban planner?" Mike asked. He was perplexed. "I can't tell them I'm an urban planner. They know I'm a laser scientist."

"No, I realize that. I'm saying I *thought* you were an urban planner. That's all I'm saying. The big thing is, we didn't talk about lasers. I have to stick with that and that's what you have to tell them."

Mike had learned the art of self-preservation; he showed little emotion. He couldn't help thinking, though, that Lewis sounded overly confident about lying to a Grand Jury; did he take his oath of office so lightly? "Well, do what you have to," he said at last. If Lewis did lie, the consequences could be serious and he saw no purpose in trying to warn him.

It was a long day for the witnesses. To Mike's relief, he was scheduled first and could escape the courthouse of clowns. It was intentional on Tuerkheimer's part. By offering Mike's testimony first, it would give the jurors an honest look at the case by which they could measure everything that followed.

Mike faced the Grand Jury for approximately forty minutes, rewinding his memory of two-and-a-half years. Prompted by Tuerkheimer's questions, he recalled his professional background and his introduction, one by one, to Claussen, then Iwen, then Lewis. It was the first time he had told anybody the whole story, from beginning to end. In a small way it was liberating. He knew he dared not wish that it was the end to anything, just the beginning of a new process, yet he couldn't help but give himself the satisfaction of a huge sigh when he left the building that day.

 Iwen and Claussen were the next two on the testimony schedule. Both pled the Fifth Amendment. They were dismissed, but ordered to provide handwriting samples. The court wouldn't press for cooperation now, knowing that it had the power to enforce it later if they saw the need.

Representative Lewis was the last of the four to testify. At 3:25 PM, Tuerkheimer began by outlining the reasons for the investigation and the possible consequences. "Mr. Lewis," he said, "Before I ask you any questions, let me advise you of the nature of the proceedings and your rights. This is a Federal Grand Jury looking into possible violations of federal law in connection with and relating to the possible illegal export of weapons. Do you understand that?"

"Yes, sir," Lewis responded.

"The investigation of the Grand Jury is at its early stage, but I hope you understand that if the Grand Jury finds at the end of the investigation that it has a reason to believe that you have committed a crime, it may indict you?"

"Yes, sir."

As he had done for the other witnesses, he concluded his introduction by explaining Lewis' rights.

Tuerkheimer began the questioning by focusing on the exhibits: letters, passport information, and travel itineraries that Lewis had provided from his files as a part of his subpoena requirements. Included was a copy of Assembly Bill 719 from 1977, which was a proposal to restrict the ownership of electronic weapons in Wisconsin. It had been designed in particular to address the stun gun, which had recently entered the market place. An amendment introduced by the Committee on Criminal Justice and Public Safety, of which Lewis was a member, had provided an exemption for laser research. Lewis was one of the three legislators who had introduced the amendment.

"Did you have any personal knowledge about the use of laser weapons at the time?" Tuerkheimer inquired.

"No, sir," Lewis answered. "My understanding is that laser weapons is [sic] light years away."

The U.S. attorney questioned Lewis' professional background and his involvement with the ultra conservative world through repeated

Code Word Tikal...163

contacts with Archibald Roberts, Albert Iwen and others. Lewis denied having any dealings with any of them apart from supporting legislation. No different than himself, Iwen supported laetrile, and Roberts had an interest in assuring that the stun gun ban didn't stifle the Second Amendment right to bear arms.

Thrown unobtrusively into the mix of who's who was Mike Muckerheide. "In what connection did you meet Mr. Muckerheide?" It was about a half-an-hour into the questioning.

Lewis recalled his first meeting with the laser scientist at Radtke Park. "I went to Wausau with Mr. Iwen. I met with Mr. Claussen and Mr. Muckerheide and we had lunch. And, we talked a little bit about my trip to, well, of Mr. Iwen's and my trip to South America and they wanted to see the pictures I took. I took about 350 slides."

"Was anything said at that time about laser technology?"

"No, sir."

Later, Tuerkheimer would ask, "Do you know what Mr. Muckerheide does, his business?"

"Mr. Muckerheide is a regional planner," Lewis answered.

"A regional planner for what?"

"I don't know. I understand he works some place in Wausau and travels about as a regional planner."

"Do you understand him to have any connection with laser systems or laser devices?"

"No, I do not."

Setting the stage, Tuerkheimer didn't press a single issue too soon, instead, diving in and out of topics before returning to lasers.

"In any of your meetings with Mr. Iwen and Mr. Muckerheide was anything said about a laser device?"

"No."

"You are sure of that?"

"I am sure of that."

"Was anything said about exporting or bringing or smuggling anything into Guatemala?"

"No."

"At any of these meetings?"

"No."

For several minutes, both Tuerkheimer and his assistant probed into the spring trip to Latin America taken by Iwen and Lewis. Lewis said that he had gone strictly to support Iwen's laetrile interests, and

On the Laser's Edge...164

during the trip had met Colonel Fuentes. When the colonel realized the state legislator's interest in photography, he offered to fly him to the Mayan ruins at Tikal on a future trip. After some correspondence, the trip was finally planned for the weekend of October 13.

The fact that a military leader in a foreign country, who was busy with an escalating civil war, would seek time to take an amateur photographer on a pleasure trip to Tikal raised some questions with the federal attorneys. So did the timing of the trip.

"It's the monsoon season in Guatemala at that time of year, isn't it, Mr. Lewis?" Attorney Judith Hawley asked.

"I don't know," Lewis answered.

Tuerkheimer would revisit the issue. "Did [Colonel Fuentes] tell you anything about weather, that you might want to watch out for weather conditions?"

Lewis claimed that he left any weather concerns to Fuentes.

Tuerkheimer persisted. "Well," he said. "Let me take a document out of the file... And as I read it, it says the rainy season generally lasts from May to November and daily showers fall during most of this period. Do you ever recall reading that?"

"As far as the photographer is concerned a rainy day is a super day to take pictures. Cloudy days and rainy days are better than sunny days."

"Do you know whether rain in Guatemala during the rainy season is a rain in which a helicopter can fly around, or planes fly, or is it heavier?"

"I don't know anything about that."

"You have no idea?"

"No. No idea at all."

One of the exhibits brought before the jury by the federal prosecutors consumed substantial attention. It was the September 6 letter from Fuentes to Lewis that Mike had intercepted for making copies. It was the letter that read:

> Dear James:
>
> Situation about our studies about "Tikal" has improved recently.
>
> I would like to know when will be possible for you to come with all your cameras and equipment.
>
> I'm very interested in our project. Please contac [sic]

Code Word Tikal...165

Mr. Iwen. Send him my regards and tell him that I want more details in order to plan a show "Tikal Show"
    Sincerely,
                Fred Fuentes

The letter was read to the jurors, followed by questions that dissected every sentence and word.

"Now, he adds on by saying, *'Please contact Mr. Iwen, send him my regards and tell him that I want more details in order to plan a show, Tikal show,'*" Tuerkheimer read. "As I understand from your earlier testimony, Mr. Iwen was not interested in Tikal?"

"That's correct," Lewis answered.

"And Mr. Iwen was not a part of any conversations between yourself and Colonel Fuentes?"

"Well, he was in the same room with me, yes, but it was just --"

"It was clear that Tikal was not one of his interests, is that right?"

"I would, just looking on the map, yes, I would say so."

"Do you have any idea why Colonel Fuentes includes Mr. Iwen in these plans relating to Tikal?"

"Well, as I understand his last paragraph, it's the fact that *'say hello to Mr. Iwen for me.'* That's all I understand. I don't understand the last sentence at all."

"Where he says, *'Tell him I want more details in order to plan a show, Tikal show,'* you don't know what he's talking about?"

"Well, I expect Tikal show would be related to the photographing of the monuments there, yes."

"Would you have any idea why the taking of photographs by yourself would be referred to as a Tikal show?"

"Well, when I get them all put together it's going to be quite a show."

".... Do you have any idea why a gentleman who participated in the *National Geographic* effort at photographing the Tikal would be interested in the photographs that you took of Tikal?"

"Well, I'm sure as a polite [gesture] that he would be very interested in the pictures..."

"...Is it possible, Mr. Lewis, that the show that Colonel Fuentes is talking about is the show relating to a laser device?"

"I don't know."

"You don't know or you're saying no?"

On the Laser's Edge...166

"I would have to say no. I don't know anything about that. Anything is possible, but as far as my knowledge of it is concerned, it's my being able to photograph Tikal."

Lewis had remained calm and professional throughout the interview. Towards the end, however, the strain was beginning to show in his answers, taking on a tart impatience.

"What happened; did anybody go down there in October?" Tuerkheimer asked him about the most recent trip planned for Latin America.

"I don't know. I didn't go."

"Why didn't you go?"

"It wasn't convenient for me to go."

"Those two weekends, those two, three-day weekends that you talked about before?"

"I went fishing."

"You preferred to go fishing?"

"I preferred to go fishing."

"I take it for a photographer such as yourself the opportunity to be flown around by Colonel Fuentes [to] the Tikal ruins --"

"He didn't call," Lewis interrupted. "The colonel didn't call and because the colonel didn't call I went fishing and I had a good time and I spent three days and Debbie's folks came up and we had a super weekend."[60]

Lewis quickly regained his composure and completed the session with his usual, political aplomb.

Two weeks later, Lewis, running uncontested, would be elected to his fourth term as a Wisconsin state assemblyman.

---

[60] Grand Jury testimony of James R. Lewis, October 25, 1978.

CHAPTER TWENTY-TWO

*Apollo sent them a fair wind, so they raised their mast
and hoisted their white sails aloft.*
— Homer, the Iliad

With some disappointment, FBI agents had gone to the Federal Grand Jury knowing that they had failed to convince John Claussen to cooperate with them. They had one more hand to deal, however, and it would be dealt on the day of testimony.

Having sought and received permission to gather handwriting samples from some of the witnesses, agents took each witness individually into an empty office near the courtroom to obtain the samples. The results could then be used by the FBI to corroborate Mike's version of the story by linking the various letters and notes in evidence to the writers. In particular, they hoped to prove that the letter to John Hartl, signed with the initials DF, for der Führer, was indeed that of Al Iwen; that Representative Lewis in fact made telephone notes regarding Colonel Fuentes; and even that Mike's written reports were made in his own hand. As for John Claussen, the FBI was interested in his alleged compilation of the Keystone Code: a simple index card that identified the various meeting places of Iwen, Claussen, Lewis and Muckerheide.

Agent Burg was conducting the procedure with Claussen. When it came his time, the investment counselor ventured into the small room in the company of attorney Joe Wiegel, took a seat at the conference table and dutifully wrote what Burg instructed. Since it is difficult for a person to disguise his or her handwriting over a period of time without revealing some inconsistencies, it was common practice for the same word to be repeated several times in an attempt to get a good sample. Burg used the contents of the Keystone Code; one, because of the psychological affects, and two, because the original code was thought to be in Claussen's handwriting and using the same words gave experts a better chance at making an identification.

On the Laser's Edge...168

Burg spoke each code about a dozen times, and each time Claussen wrote it down, making no comment.

"B-1, Hoffman House...B-4, Howard Johnson's...S-5, Sambo's..."

Burg held his trump card for last. When he came to the word *Apollo*, instead of a dozen times, he repeated it over and over and over again.

At first mention of Apollo, Burg thought he noticed some hesitation, but Claussen obligingly wrote down the word.

"Apollo," he repeated.

The same response.

"Apollo... Apollo.... Apollo..." the agent continued, his voice carrying the word on a strong and decisive course. "Apollo.... Apollo.... Apollo."

Finally, after nearly two-dozen recitations, Claussen broke. "How did you know about that?" he asked nervously, and Burg knew he had made a breakthrough.

"Do yourself a favor," he advised Claussen in Wiegel's presence. "Hire your own attorney."

The next day, a new attorney, Frank Bachhuber, Jr., called the Wausau FBI office and began making arrangements for Claussen's cooperation. That Saturday, Agents Burg and Szekely met with Claussen at Bachhuber's office in Wausau to set things in motion.[61] In a twist of irony, the man who had gotten Mike involved in the laser conspiracy to begin with, was now the one poised to help get him out.

In the months ahead, additional testimony from Claussen would be provided to the Grand Jury in a sworn, signed statement. Though varying in small details to Mike's recollection, the investment advisor admitted that he had been the one to orchestrate the initial meeting between the chemist and the laser scientist because of their mutual, political conservatism and Mike's curiosity about laetrile. He recalled meeting Iwen first in 1972 at a political meeting, then Mike sometime during the next two years in the course of his investment work. The day he brought Iwen and Mike together for the first time at the George Wallace rally, he recalled Iwen saying to Mike, "Would you like to work for me sometime? I have a job for you." It was not long before it became clear that he was referring to the development of laser weapons.

Claussen noted that Iwen's interest in lasers began to escalate after his first trip to Guatemala when he connected with foreign offi-

---

[61] Interview with FBI agent Tom Burg.

cials. "He stated that it was his desire to mass produce lasers by setting up a plant in South America. These devices were to be antipersonnel laser weapons...I determined his intent was to have soldiers of the governments of foreign countries equipped with these rifles," Claussen said.

He agreed that Lewis arrived on the scene in 1978, although his name first surfaced in conversations the previous year. Of Lewis' contention that he had no knowledge of laser weapons Claussen said, "I was present at several meetings with James R. Lewis during which there was discussion of arms and laser weapons. Lewis was present when Mike Muckerheide exhibited his laser demonstration materials in Stevens Point, Wisconsin, in June 1978. Any statement made by James R. Lewis to the contrary is false."

According to Claussen, both Iwen and Lewis shared a vision for change. "Iwen stated his belief of an international conspiracy. [He] stated that there was a building in Chicago, some sort of an international building or office, which he wanted to blow off to the foundation with a laser. At this first meeting, Lewis made the statement 'What I'd like to do is take a plane and fly by the control tower of O'Hare Airport and beam a laser on the tower and blow the whole thing up.' He said he would then 'Beat the country (get out),' as he and Iwen felt that there was a nuclear war coming."

Cuba was another target of mutual interest. "Both Iwen and Lewis have stated the desire to 'fly over Cuba in a plane and hit Havana Harbor with a laser beam.' Iwen dislikes Communism and Castro and this was the reason behind the idea," Claussen noted.

As much as they might have wanted to use such a destructive force themselves, Claussen said they opted for directing their energies towards the path of producing and selling weapons to a foreign government. They kept their evolving plans under wraps by using the word *Tikal* as a code word for the project. "I am aware that it relates to Indian ruins in Guatemala," he said. "Iwen had talked of Indian ruins before, and this was partly legitimate, as James Lewis had talked about photographing the ruins in Guatemala. However, Iwen once mentioned that this would be a cover for other operations..."

He remembered the letter from Colonel Fuentes mentioning Tikal. Lewis had said in his testimony that the colonel was only referring to the Wisconsin assemblyman's plans to photograph the Mayan ruins. Claussen had a different interpretation. "In the letter was men-

On the Laser's Edge...170

tion of Tikal, which I recognized from its usage to be a code."

As hoped, Claussen's testimony exonerated Mike. He was painted as a cautious man who had always remained skeptical. "Muckerheide was evasive," he noted, "and to my belief, never produced any hardware, nor gave Iwen any literature other than published articles." Sometimes, he added, Muckerheide "appeared scared" at some of Iwen's ideas, particularly at the destruction of buildings in the U.S. "I don't know if Muckerheide went along with this idea or not," he admitted. "Sometimes he would say, 'I could make a machine to do this,' but I was never sure if he was serious."

As for his own role, not even Claussen really seemed to understand it. "I was there really because I was interested in hearing Iwen's political talks," he contended. He didn't always take what was said seriously, but his desire to be included in the group was clear. "I once showed a pistol in my car to Mike Muckerheide to put on a front or create an impression with him." [62]

When it came right down to it, Claussen was a man who simply wanted to belong.

His statements were a welcome bonus for the prosecution. The original Grand Jury members were called back to their posts in early 1979 to see the case through, despite the fact that a new Grand Jury was already seated. The investigation would continue to consume most of that year.

With Claussen's ties now severed, and Lewis' attention distracted by the investigation, it was easy for Mike to distance himself from the group. With each of them now represented by separate attorneys, Mike would simply tell them he had been advised not to make any contact. It was hardly necessary since the four men were quickly losing interest in maintaining their ties. Even Iwen, who had advocated for solidarity, withdrew quickly.

Mike's last significant contact with Iwen was in March of 1979. Iwen called him several times to ask that he return the gun that Lewis had given him. He never gave him a good reason for wanting it, but Mike began to pick up on the likely reason through other contacts. President Carter was scheduled to visit Wisconsin on March 31 and there were rumors that the Posse Comitatus planned to assassinate him. Lewis, who heard the rumors, was afraid that the gun he had

---

[62] Sworn Statement of John Claussen taken by Agent Burg on February 8, 1979 and presented to the Grand Jury as Burg Exhibit #1 on May 2, 1979.

given Mike might be used in the attempt. The gun had Lewis' fingerprints all over it and he thought the authorities might try to frame him for the crime. He wanted Iwen to get it back.[63]

The Secret Service investigated the rumors, talking to several likely suspects prior to March 31. President Carter came and went from Wisconsin without incident. The gun remained in the safety deposit box until retrieved by the FBI later that year and ultimately put into an exhibit used for bank employee seminars.

Liberated from the daily diversion of the laser plot, Mike began to spend even more of his time in the research laboratory, with impressive results. That year, 1979, he received the President's Award from the Wisconsin Association of Vocational Adult Educators for contributions to the advancement of practical education, and he saw two more of his patents accepted after years of work and waiting. The first was his co-invention with Dr. Uecker and Dr. Mallozzi's team from Battelle Memorial Institute for a method of applying radiation at a desired location using laser-generated plasmas as a source of x-rays, filed in 1977. Financed in part by the North Central Health Foundation, which was in the process of dissolving, the patent was assigned to Battelle. The second was a patent for his invention of a laser system for vaporizing and absorbing material at inaccessible areas, filed in August of 1978. Mike was beginning to widen his contacts and his reputation beyond Wausau and Wisconsin, becoming affiliated with some of the best laser minds in the U.S. New ideas for patents were surfacing in his research.

The majority of his work was in the field of medicine. To prove to himself and others that he could develop the kind of blinding rifle that Iwen had pursued, he used his spare time to build one. That spring, he developed a device that was capable of both temporary and permanent eye damage that he thought might ultimately be of interest to law enforcement. The gun was submitted to the Food & Drug Administration to be tested by their Electro-Optics Specialist. It was never put into production.[64]

---

[63] "Jim Lewis feared his gun would be used to shoot Carter," by Jack Anderson, *West Bend News*, November 21, 1980.

[64] The reality of weapons designed to blind, temporarily or permanently, a human target was unknown by the general public in the 1970s. By 1995 the technology was the focus of controversy over how humane it was. In 1998, an international ban went into effect on weapons designed to blind.

On the Laser's Edge...172

As for the case, it dragged on month after month. A constant source of worry was the lingering uncertainty of the Grand Jury probe and the fear that Mike's role would be revealed at an unexpected time. He was never sure when or if any charges would be filed, and was never comfortable that his family was being adequately protected.

He turned more paranoid. While at North Central Technical Institute one day, he and one of the soon-to-be first graduates of the school's laser program, Dan Szygelski, were walking together towards the cafeteria for a cup of coffee. Suddenly, someone ran towards them from the opposite side of the crowded room, snapped a picture and raced away. It happened so quickly that Szygelski couldn't even tell if the photographer was male or female. Both men had no doubts, however, that one or both of them were the intended target. Mike was extremely upset. "He apologized and said, 'I gotta' go,'" Szygelski recalled, and rushed out, leaving the student totally bewildered.

Other people working around Mike also noticed the difference. Normally pleasant and amusing, there were days when he would plunge into dark and serious moods. In his tiny office at A. Ward Ford Institute, where the walls were plastered with notes and diagrams, he would lock himself away for hours on end. Mysterious phone calls were his only connections to the outside world. Sometimes, one of the Institute's young employees, Diane Dalsky, would be taken aside and asked to take a message. "He would tell me who he was talking to and I would have to verify that he actually had a conversation with that person, and sign my name, the date and time." She said she did not recognize the names, and never discovered exactly what it all meant.

Concerned for the safety of the people he worked with, along with his family, friends and neighbors, Mike began to drop ominous hints about his secret life as a precaution to safety. "It's very serious and things could happen," he would tell them. Sometimes he would even reveal some of the conversations and episodes: his trips to West Virginia and seeing the underground laboratory; the plans to trigger a catastrophe at O'Hare Airport and Havana Harbor. It all sounded too odd and unbelievable and, as with the FBI initially, Mike encountered another round of skepticism. This time it came with a twist of irony. The FBI had understood the politics of extremism, but not the laser science. Now, Mike faced people who understood the science but not the bizarre nature of political extremism. It all came down to the same

Code Word Tikal...173

thing: people were sure he was a scientist going mad.

At least one of the people Mike called frequently during these long and agonizing days of the Grand Jury probe was Agent Burg whose belief in him never wavered. It was easy talking to him, and Burg always knew what to say. Typically it wasn't much, and often it was the same thing Mike had heard before, but it kept him grounded.

"How can I protect my family?" Mike often worried. "How can I make sure nothing happens to them?"

Burg wouldn't try to put a happy face on his concerns, he would just tell him the truth. "There's a certain degree of risk," he would admit. "There always is. But your role as our source is still unknown to these guys, and because everyone has been advised to stay away from each other, I don't think there's much of a chance they're going to find out prematurely.

"To be on the safe side, though, you and your wife have to stay alert to anything suspicious. Drive your daughter to wherever she needs to go. Know the kids she plays with and know their parents.

"On the good side, you live in a quiet neighborhood where people look out for one another. My experience says nothing will happen. These guys don't want to do anything stupid now. It would just complicate their problems. They know that."

After these reassurances, Mike would feel relieved for another few days; go back to his work and his life. Slowly the doubts would creep in again and he would call Burg, reaching out for his lifeline to sanity.

It was during this time that Burg dared to visit the Muckerheide home one day. He was dropping Mike off after the two men had made a trip to Madison for more Grand Jury testimony. They had spent the trip there and back talking at length about everything imaginable: the case, careers, laser science, family life. When they reached the Muckerheide home, Mike asked Agent Burg to take a moment to meet his family.

Mike's daughter, then 13-years-old, was in the backyard, playing. It was a neat little backyard: a place where Mike liked to spend his few moments of spare time. There, several winters earlier when his career was less demanding, and while other fathers in the neighborhood were building snowmen and forts, Mike spent more than eight hours building a full-scale locomotive, complete with a smoke stack belching steam from a block of dry ice. In the same yard, he had once built a wooden, gingerbread house for his own little Cinderella, with a holographic mirror that held the image of the evil witch.[65]

Readily, Susan came over to her father, smiling shyly at the stranger. She was developing the good looks and subtle charms of her mother.

"I want you to meet Mr. Burg," Mike told her. "What he does is important. The fact that you can play in your own backyard, free and unafraid, is an important piece of your America, and men like Mr. Burg are protecting it. Never forget that. There are people out there that work to protect what is good."[66]

Burg was flattered, but left wondering if he could live up to Mike's expectations. There was some information he hadn't yet shared with him, primarily his own frustrations over where the case was — and wasn't — headed.

---

[65] Interview with Janet Volpe.
[66] Interview with Mike Muckerheide.

CHAPTER TWENTY-THREE

Let me enjoy the earth no less because the all-ending light that fashioned forth its loveliness had other aims than my delight.
— *Thomas Hardy*

Agent Burg needed to obtain information from the government about laser weapons. He needed to give the Grand Jury an idea of how feasible and dangerous they could be; to enlighten the jurors to the fact that laser weapons were no longer the products of a writer's imagination; they existed.

Agent Burg, however, was hitting a wall. The people in charge of making the weapons weren't willing to talk. The information he needed was highly classified and confidential.

He contacted experts at Aberdeen Proving Ground in Maryland, the Department of Defense in Washington D.C., and electro-optic specialists at Honeywell and the Food and Drug Administration. They all agreed that lasers were of some interest to the military, and that certain lasers were capable of blinding people, but they downplayed their potential. One military expert even claimed that he had once looked into the beam of a powerful argon laser without being blinded. By downplaying the seriousness of lasers, Burg believed the official was trying to discourage any charges that might arise and shed too much public light on the subject.

Burg had to be resourceful. He latched onto any published materials he could find, including a March 26, 1979 issue of *Aviation Week* and *Space Technology* which reported that the Carter administration had blocked potential military sales of laser-lok sights to Guatemala. It wasn't much, but it went into the file as further evidence that military lasers were being developed, and bringing with it all the baggage of determining who should have access to it and how it should be handled.

The unwillingness of government experts to reveal anything substantial about the possibilities for laser weapons helped resurrect the same doubts that had plagued the case from its beginning. It started in

On the Laser's Edge...176

Milwaukee, where Burg discovered that there existed a real fear in pursuing charges. There were multiple reasons why. First, without the military evidence, it was difficult for some people to imagine the dangerous side of lasers. Second, Central Wisconsin was known for contented dairy cows, not criminals. Its residents were quiet and hardworking, its outlaws confined to the larger cities. Even within the Bureau, where everyone was trained to be skeptical, it was a stretch to believe that such sophisticated scheming could be born and bred in the state's heartland. And third, the Milwaukee field office could end up as the laughing stock of the Bureau. The corruption of a politician was one thing, but a conspiracy about making and selling blinding laser weapons to a foreign government, or to install one into an airplane, just seemed too ludicrous.

Eventually, Supervisor Byrne would take a hands-off approach, "staying at arm's length and hoping that things didn't blow up in Wausau and splash all over him," Burg described. The case was passed up through the ranks from one supervisor to the next, each one more skeptical than the next. It finally landed on the desk of Unit Chief Jim Graham at the D.C. Headquarters, where it floundered. On Burg's attempts to breathe life back into it, he found himself caught in a circle of excuses. Byrne would tell him that he knew nothing about the case because he had passed it up to Graham, and Graham would refuse to talk to him because protocol dictated that he deal with supervisors, not agents. The only hope lay with Tuerkheimer.

Things were still moving along in Madison with some momentum. The Grand Jury called back some of the witnesses for additional testimony, and by midyear of 1979 the jurors were convinced that the evidence was more than sufficient to give Tuerkheimer the green light to bring charges against Lewis and Iwen. Because existing regulations in foreign commerce did not specifically regulate the cutting edge technology of lasers, Tuerkheimer was preparing to use a patchwork of laws, bringing it all together under a charge of conspiracy to defraud the government of its rightful role in conducting foreign policy.

Tuerkheimer knew that the charge would be a challenge to prove. Prevention of a crime can be a curse to the prosecution and, in this case, the most serious crimes had been prevented. The fact that Lewis had CIA names in his possession did not prove that he intended to pass them to anyone. The fact that Iwen and Lewis were discussing the sale of weapons to a foreign government did not mean that they would actually go through with it, especially since Mike had avoided making the actual weapon. Still, he believed, as did the Grand Jury, that all the evidence together: the FBI's interrogation of suspects, the years of information provided by Mike, Claussen's testimony, and Lewis' taped interview, were more than enough to win convictions.

Prior to bringing charges, Tuerkheimer contacted Lewis' attorney to outline what was about to happen. He would reconsider his course of action if he obtained the assemblyman's cooperation. Between Iwen and Lewis, Lewis stood to lose the most by a conviction: his reputation and his elected office. The pressure was on him to make the next move.

CHAPTER TWENTY-FOUR

*In war, you can only be killed once,
but in politics, many times.*
— Winston Churchill

As the calendar turned to August, it also turned a new page in the laser case. Lewis' attorney, Francis R. Croak of Milwaukee, had notified Tuerkheimer the month before that his client was willing to cooperate. Lewis would plead guilty for lying to a Federal Grand Jury in exchange for not having to face additional charges in the laser conspiracy.

On hearing word of the decision, Burg was elated. First Claussen had folded, followed by Lewis. Two down: Iwen to go.

He called Mike to give him the good news, with a cautionary note. "Mike, I just want to make sure that you're all right with this. Tuerkheimer really needs you to help hold this case together. It would mean using your name and all the information you gathered. It's going to be out in the open now."

Mike didn't have to think about it. He had been down that mental road so many times he just wanted it over with and behind him. Another year had passed without justice or resolution. "It's all right," he said. "I'm okay with it. I just want you to get these guys. They've put me through hell."

In FBI terminology today, Mike's willingness to reveal his identity in the case would have ended his role as an Informant and reclassified him as a Cooperating Witness. To Burg, the term Confidential Source was always more appropriate because Mike had always been willing to ultimately walk out of his veil of secrecy. By any name, he was in a dangerous position, not knowing how Iwen, Lewis, Claussen, or any of their sympathizers, might react to the news.

On the afternoon of Monday, August 27, 1979, Lewis walked into the federal courtroom in Madison to enter his guilty plea. It was no surprise that none of Lewis' cohorts were there for the occasion, not even Mike; but, it was a surprise to see who was there: every one of the grand jurors who had reviewed the evidence. They were not there because they had to be; in fact, they did not. They were there because they felt that they were about to witness something very unusual. "This is the big one," one juror was overhead saying, summing up a universal thought that this court appearance would be the beginning of long and intriguing journey to the truth.

Federal Judge James Doyle, Sr., whose son Jim would become Wisconsin's governor in 2003, presided over the court that day. Everyone was waiting for him so the proceedings could begin.

First to arrive through a side door was the court reporter, followed by the court clerk. Almost simultaneously the bailiff called "All rise" and the judge entered through a separate door behind his bench. The spectators seemed confused as they flipped their attention from the court clerk to the judge, the reporters knowing they should train their focus on the judge. The men in the room, however, lapsed easily back to watching the clerk, Cathy Fahey. Even as she tried to enter discreetly, the tall and statuesque Fahey commanded notice. Heads pivoted, conversations lulled until she had taken her seat.

After Judge Doyle, and Fahey, settled in, Doyle immediately began. "I do accept the agreement that has been reached between the prosecutor and the defense and that is embodied in Court Exhibit 1," he said, "And it will be honored in any further proceedings in this case.

Mr. Tuerkheimer, would you please state what the Government would be prepared to prove if this case were to be tried?"[67]

Tuerkheimer stood to make his case, not losing sight of the irony. Many times he had stood in the same courtroom watching as waves of eager immigrants took the oath of U.S. citizenship. "There would be hundreds of people there sworn into citizenship," he recalled, "...and each one, when they were sworn in, knew it was something important... This represented to them a culmination of what they'd wanted, so they brought an atmosphere of reverence and hope..."[68]

In that same courtroom, on this day, Tuerkheimer was watching the flip side: an American born citizen, elected to serve the people, falling in disgrace for lying in a case bordering on treason.

The federal attorney cited the letter from Colonel Fuentes as his key piece of evidence on which the charge would be based, along with

---

[67] *United States versus James R. Lewis*, U.S. District Court, Western District of Wisconsin, case #79-CR-57, Lewis plea, August 27, 1979.

[68] Interview with Frank Tuerkheimer.

On the Laser's Edge...180

Lewis' false testimony given to the Grand Jury. Citing evidence to the contrary, he drove home the point that Lewis knew from the beginning that Fuentes wasn't referring to a photo expedition to Tikal, but rather a laser show. Not only that, but contrary to his testimony, Lewis had been an active participant and proponent of the illegal weapons plan. His third blunder was contending that Mike was an urban planner when it was obvious that he knew otherwise.

For the first time, Mike's identity as an FBI source was uncovered to the public.

"In June of 1978," Tuerkheimer said, "Mr. Iwen, Mr. Lewis and others met with a laser scientist by the name of Myron Muckerheide...Mr. Muckerheide had, in the past, been contacted by Mr. Iwen; and when he had suspected that something might be wrong, he immediately provided the information to the Federal Bureau of Investigation; and that set a pattern which was to continue and which was in fact continuing at the time that Mr. Lewis and Mr. Iwen were speaking to Mr. Muckerheide in June of 1978..."

"Mr. Lewis, I'd like to ask you directly whether the facts are basically as the United States Attorney has just stated them?" Judge Doyle asked when Tuerkheimer had finished.

"This is correct," Lewis responded.

"Thank you.

"On the basis of this discussion with the defendant and his counsel, I find and conclude that the defendant has knowingly and understandingly and voluntarily entered a plea of guilty to the information charged against him today by the United States Attorney, and that he has done so after an adequate opportunity to consult with his counsel, with an understanding of the nature of the charge, and with an understanding of the consequences of the plea.

"I accept the plea. I also determine that there is a factual basis for the plea. And on the basis of the plea, I do adjudge the defendant guilty of the offense charged in this Information..."

Judge Doyle ordered a 30-day pre-sentence investigation, and the court was adjourned. Representative James R. Lewis now faced an uncertain future that could mean a maximum sentence of five years in prison and a fine of $10,000.[69]

---

[69] *United States versus James R. Lewis*, U.S. District Court, Western District of Wisconsin, case #79-CR-57, Lewis plea, August 27, 1979.

News of the guilty plea spread quickly. Among the first to hear it were FBI agents and police officers attending a Sex Crimes Police School at the Holiday Inn on the east side of Madison. Agent Burg was scheduled to be a speaker at the school, but was set to make a later presentation in order to attend the court proceedings. By the time he arrived at the hotel, the radio and TV media had already announced the news. The officers and agents attending the school greeted Burg with handshakes and congratulations that extended into the night. It was a major case and one of the few involving a Wisconsin official. Burg couldn't stop the flow of free drinks that came his way as they gathered at the hotel pub that evening. Appropriately, Agent Southworth was there for the occasion. He had decided he did not like working at "Disneyland East," (the nickname given FBI Headquarters in Washington, D.C.), and was back in Wisconsin working at the Eau Claire Resident Agency.

Until that day, despite widespread rumors, the Grand Jury investigation had remained out of the public eye; its testimony held, according to law, in strict confidence. The plea by Lewis thrust the case into daylight, and news reporters from throughout the state pounced on the story.

The headlines blared the assemblyman's guilt, and speculated on his future. Since a convicted felon couldn't serve in public office, it was expected that he would serve jail time and lose his seat in the State Assembly.

The future for Mike was less certain. With the bulk of the background details still locked away in confidential Grand Jury testimony, the media could do little to explain what had led to Lewis' guilty plea, only vaguely outlining a conspiracy involving a couple of scientists and lasers. Because Mike had agreed, for the sake of the case, to reveal his identity, he was left to deal with a public who only knew part of the truth, and a very bizarre part of the truth at that.

He was not the type to worry about what people thought, however. In fact, he was confident that the full truth would see the light of day when additional charges were filed and justice meted out. He only worried about his family. How could he protect them from criticism and from someone who might want to harm them physically? Buoyed by their strong religious faith, the family chose to live their lives as normally as possible, facing the crisis head-on.

Lewis' sentencing was set for November 8, but he started early to back pedal for leniency. He latched onto a double-agent theory, claiming that he had only feigned his role in the laser scheme to collect information he planned to turn over to the proper authorities.[70] Judge Doyle delayed the sentencing to look into the matter and found the idea "preposterous."[71] The sentencing was rescheduled for November 21, the day before Thanksgiving.

In the meantime, Lewis went about his work as usual, meeting as many of his daily obligations as possible. With so many of the facts of the case still unknown, fellow assemblymen from both political parties rallied to his side. A leading Democrat gave him an informal cocktail party soon after he entered his guilty plea. Leaders from both parties in the Assembly attended. His colleagues described him to the media as "the most principled conservative in the Legislature," "a sincere, dedicated, hardworking legislator," and "articulate and courageous."[72] One legislator was quoted as saying that the FBI "could have better spent their time looking into the heavy drug traffic, which every day is sending young girls into prostitution, young boys into stealing and lots of people into institutions."[73]

Despite his sometimes ultraconservative viewpoints, Lewis had built a good reputation among both Republican and Democratic colleagues in the Assembly. In his seven years in Madison most of his bills dealt with issues of law and order. While some of them were politically charged, like his support for reinstating the death penalty, legalizing the medicinal use of marijuana, the legalization and control of laetrile, and the recording of all abortions as deaths, he also fought hard for more mainstream legislation such as bills against pornography and for prohibiting convicted felons from carrying dangerous weapons. He was also a proponent of a decidedly more liberal notion of prison reform.

Just prior to making his guilty plea Lewis had garnered so much

---

[70] *United States versus James R. Lewis*, U.S. District Court, Western District of Wisconsin, case #79-CR-57

[71] Details of the sentencing taken primarily from the article, "Lewis Gets 6-month Term" by Jack Anderson, *West Bend News*, November 21, 1979.

[72] "Lewis and lasers: why?" by Jack Anderson, *West Bend News*, August 29, 1979.

[73] "State legislator pleads guilty to felony charge" by Associated Press, *Wausau Daily Herald*, August 28, 1979.

support that he was being considered for a top-level position in the corrections department.

He had charmed constituents, colleagues, and, supposedly more than his share of women, but could he charm his way out of prison?

At 9:15 AM on the day of his sentencing, Lewis walked into the courtroom wearing the pleasant smile of confidence he was known for. He was, as always, impeccably dressed and groomed. At his side was his former girlfriend, now his wife, Deborah Batzler Lewis, whom he'd wed in February.

Lewis' attorney, Francis Croak, was given the first opportunity to address the court. In his mid 40s, the tall, solidly built Croak set a low-key and dignified tone. For the next thirteen minutes he made his final appeal to seek mercy for his client, explaining Lewis' many years of loyal service to the state.

He pointed out that Lewis had no previous court record, instead showing "a lack of anything other than a complete adherence to the law." He had received many letters of support that clearly showed "a genuine respect from people of different political persuasions" and asked for the court's mercy for a man who would already suffer greatly by the loss of his right to hold public office.

U.S. Attorney Tuerkheimer would follow Croak, painting a less flattering picture of the tarnished assemblyman. He said the letters of support for Lewis were "laudatory" but asked, "What is the significance of other people's opinion of Lewis?"

The evidence showed a different side of the elected official than what he had revealed to his constituents and his colleagues. "He was using his office, [Senator] Proxmire's office and [official government] letterhead for what Lewis later admits is private business. It's clear it's a cover-up," Tuerkheimer told the court, adding, "This is a man willing to use his office [for personal reasons]...and with complete disdain." It was a measure, he said, "more revealing of what his character really is" than any letters of support.

Last to speak before the sentencing was Lewis. He and his attorney stood together, facing the judge. Softly and slowly he spoke: "I sincerely apologize to the Legislature and to the people I represent. I recognize the shame I brought to the body I was entrusted to serve in. I'm very sorry for those circumstances that brought me here today. I have nothing more to say."

As always, the final word came from the black-robed Judge

Doyle. Decisively he began: "Although it should not be necessary to say it, I state explicitly that it would be a grave abuse of the power of this court if its sentence were to be influenced in the slightest degree by the nature of Mr. Lewis' beliefs on public issues or by the nature of his participation in the political process."

He conceded that Lewis was not the prime mover in the plan to take laser devices to Guatemala. "The hatching of the scheme had preceded his awareness of it," he noted. At the same time, "...it was not his purpose to simply obtain information about the scheme in order to turn it over to the proper law enforcement authorities. His role lay somewhere between these extremes. His involvement was heavy and it persisted over a period of time."

The judge said it wasn't clear why Lewis had lied to the Grand Jury, "whether it was simply to save himself or to protect the scheme and the schemers," but most likely to keep his own role in the scheme undiscovered.

"Obviously the offense is grave," he added. "Law and order were subverted by answers which impeded an investigation into criminal conduct.

"I believe that [Lewis'] experience, without any sentence from the court, has been harsh enough to deter [him] from any further criminal conduct. But others must be shown that if they choose to obstruct justice in this way, the penalty will be severe, whether the offender is an obscure person or a prominent public figure.

"Because of his public service, which has won for him the respect of his fellow legislators of all political persuasions, I have limited the penalty in Mr. Lewis' case to the minimum I think necessary to send the word to others."

The sentence was pronounced just before 10:00 AM: six months in prison, 30-days less for good behavior. Lewis was ordered to surrender to the U.S. Marshal on December 3, 1979 to start serving his term. The Federal Bureau of Prisons would later decide that he would be sent to Camp Leavenworth, the minimum security facility at Leavenworth, Kansas.

As Doyle announced the sentence, the only notable sounds in the room of thirty spectators were the quiet sobs of Deborah Lewis, and the scratching of hurried pens on news media ledgers. As the court adjourned, everyone filed out of the courtroom except Lewis, his wife, and his attorney. They would stay for another twenty minutes.

Waiting for them in the hallway were a dozen news reporters. Lewis would make his way out of the room and down the hall through a gauntlet of questions. He smiled and made no comment except: "I'll say, have a happy Thanksgiving, and praise the Lord for that."[74]

---

[74] The recounting of the Lewis sentencing was taken primarily from the article "Lewis Gets 6-month Term" by Jack Anderson, *West Bend News*, November 21, 1979.

CHAPTER TWENTY-FIVE

Thrusted light is worse than presented pistols.
— *Herman Melville*

Every day in our world there are countless people who live on the sharp edge between crime and justice. They risk ridicule, and even life, to be there, and often take no credit, are offered no credit, for the crimes they've prevented.

Mike was one of those people who lived on that edge. He looked forward to the day when he could live a normal life, to feel that everything he had sacrificed had been worth it.

With Lewis on his way to prison for perjury, Mike waited anxiously to see what additional charges would be filed. Tuerkheimer was looking at two people in particular: Iwen and possibly Fuentes, both for illegally dealing in arms.

On the day of Lewis' sentencing, Tuerkheimer made his announcement to the media: "The investigation is over. The case is closed."

Nobody saw it coming, especially not Mike.

"I'm sorry. I had no choice," Tuerkheimer explained to Agent Burg. There was a hint of frustration in his tone, but Burg was limited in how much he dared pressure him for a reason. Burg was now heavily involved with another headline-grabbing case; one involving a State Circuit Court judge in northern Wisconsin who was operating a prostitution ring. He couldn't jeopardize the good working relationship he had developed with the U.S. Attorney over something that couldn't be changed. "By now, I really liked working with Frank," Burg admitted. "I had discovered that 'that goddam Tuerkheimer' was actually a great guy."

The part Burg dreaded was trying to explain the decision to Mike. He knew it would cut into his soul as deeply as one of his lasers.

"Hey, it's great, though, that we got Lewis," Burg tried to console him when he met Mike in his office four days after the Lewis sentencing. "He could have recanted and maybe gotten away with it."

Mike wouldn't buy it. To begin with, he wasn't happy that Lewis was only getting six months in prison. It was before the era of Determinate Sentences and Federal Sentencing Guidelines, which began in 1988, so no one had really known what to expect. Mike had certainly expected more.

"My God, what's six months after all the years and all the crap I've been through?" he complained to Burg.

"Mike, you don't know how lucky we were to get the six months. Doyle is a lenient judge. It's a first offense and usually that means probation. I think he was sending a clear message that he took this very seriously. Besides, Lewis isn't just losing a few months of a normal life. He's lost his job. He's lost his reputation. He'll never hold public office again."

So, what about Iwen, Mike wondered, thinking about all the laws that had been broken over the course of three years: transportation of loaded and concealed weapons, threats to bodily harm, not to mention the larger issues of fraud and conspiracy. He could not and would not ever understand how Iwen could get away clean. He shook his head in dismay. "Something is really wrong when people like Iwen get away with it," he said. He felt betrayed by Tuerkheimer.

Burg had always been able to find a way to get through to Mike and to give him some words of encouragement. This time, he knew he had failed. As similar as they were, he and Mike viewed the decision very differently. To an agent, success was based on the ability to thwart crime, and in this case the laser conspiracy had been successfully uncovered and stopped. Right or wrong, it didn't always lead to charges. Agent Burg accepted that as part of his job.

Mike, on the other hand, had no job description that could help him put the case in perspective. He believed that for every action there was a reaction; for every wrong, a consequence that had to be paid. It was why he had gone to the FBI back in 1976. When someone threatened to harm people, to harm his country's freedoms, shouldn't they be caught and punished?

"Trust me," Burg said, "Iwen's about to stumble again, and we'll be there when he falls."

It wasn't much consolation and Mike would call Burg several more times in the days ahead still hoping for a different answer. Finally, Burg arranged for him to speak to the Special Agent in Charge. Hogan had retired, but taking his place was a rising star, John

On the Laser's Edge...188

D. Glover. He was the first black SAC in the Bureau and would one day become the third highest ranking man in the FBI.

Mike would meet with Glover in early January 1979 in Milwaukee. From his office overlooking the heart of the snow-covered city, Glover would set the story straight. It wasn't Tuerkheimer who had tipped the case. The decision had gone over his head, all the way up the chain of command to the higher echelons of the Justice Department in Washington. Tuerkheimer had been told to leave the laser case alone.

Like Burg, Glover promised Mike that Iwen wouldn't get far. "Tom's got Iwen in the cross-hairs," he said, leaving Mike to ponder his meaning.

It would be years later before Tuerkheimer would explain in more detail about what had happened to the laser case and what valuable lesson it taught him about the U.S. justice system. By then, he had left his federal position to enter comfortably into academia and private practice.

Heading into the final phases of the Lewis case, Tuerkheimer explained, he had every reason to believe that it was only the beginning of more charges ahead. Then, in the eleventh hour, he received a phone call from a top official at the Justice Department in Washington, D.C. "What the hell are you doing?" was the basic message. The call changed everything.

Throughout his career in the federal justice system, Tuerkheimer had enjoyed a great deal of freedom. He had gained his experience in the Southern District of New York, which had a history of independence. He knew the law, he had the confidence to prosecute it, and he was left alone to make the necessary decisions.

Not this time.

"Read your Justice Manual," he was told. "In matters of national security, *we* decide if there's going to be charges filed. You do *nothing* until we look at the information first."

Tuerkheimer realized then how close to reality were the laser weapons of Iwen's dreams, and how possible it would have been for someone with Mike's knowledge to create them. On the national level, such knowledge had become so important that it had been sealed off in a classified world. If a trial were held, the technology of laser weaponry would have been forced into the open, possibly compromising secret U.S. military research and development. If Fuentes had been dragged into the fray, it might have jeopardized the United State's delicate relations with Guatemala.

It was an unpleasant wake-up call for the confident U.S. attorney. He apologized for his oversight and vowed never to make the same mistake again, knowing full well it would cost him his job if he did. "Getting an Iwen was not worth it," Tuerkheimer decided. He sent the case off to Washington, D.C. expecting never to see it again.[75]

At the time the case was shut down, Agent Burg had been privy to some of the reasons behind it. As an agent, who also answered to a higher office in the nation's capitol, he understood it; yet, he did not like it. He told Tuerkheimer that his own opinion of the Justice Department plummeted as a result of the laser case to "somewhere in the basement" of his opinion, to which Tuerkheimer responded, "There are subbasements you can't imagine."

---

[75] Interview with Frank Tuerkheimer.

CHAPTER TWENTY-SIX

The big thieves hang the little ones.
  – *Czech Proverb*

Twelve days before Lewis was sentenced, 24-year-old Kevin Buggy walked into a shopping center in Brookfield, Wisconsin and paid for his merchandise with $20 bills. They were counterfeit.

Buggy was arrested by the Brookfield Police and questioned by federal authorities, including the Secret Service. They wanted to know where he had gotten the money.

The information they gathered from Buggy worked its way up through the system to U.S. Attorney Joan Kessler. Kessler saw something particularly interesting about the case, and passed it on to Agent Burg. Burg was elated. The counterfeit trail was leading to Albert C. Iwen.

According to Buggy's testimony, Buggy and Iwen met at the V.M. Nutri Company in Lake Geneva where Iwen was a consultant and Buggy an employee. Iwen approached him for a $2,000 loan. When Buggy asked to be repaid, Iwen offered him the counterfeit bills that he claimed were being manufactured by the Posse Comitatus. Buggy agreed to take the money, then spent it at Brookfield Square.

Counterfeiting was something solid on which the government could build charges against Iwen. It wasn't illusive like the manufacturing of a controversial drug, or dealings over a laser weapon that had never been built. They were anxious to offer Buggy a deal: we'll go easy on you if you help us get Iwen.

Buggy's contribution would be two, secretly taped interviews with the chemist. Ironically, Iwen warned him to be very careful about whom he trusted. "I've had experience with that sort of thing," he forewarned Buggy. "Guys acting...working for the Feds. Entrapment schemes. In a laser caper...up north."

On November 30th, eight days after the Lewis sentencing, Iwen pulled his car up to a drive-through window at a McDonald's

Restaurant in Fond du Lac, Wisconsin handed the clerk a counterfeit $20 bill and sealed his own fate. He wouldn't be arrested for another three months, but by then the government had more than enough evidence for a conviction. On July 20, 1980 he was sentenced in U.S. District Court, Milwaukee for three counts of counterfeiting. He served three years in prison.[76]

---

[76] Eastern District of Wisconsin, Court case #80 CR 35, *United States versus Albert Carl Iwen*.

CHAPTER TWENTY-SEVEN

Dwell not upon thy weariness,
thy strength shall be according to the measure of thy desire.
– Arab proverb

Wisconsin Assembly Speaker Edward Jackamonis, a Democrat from Waukesha, ordered Lewis' name removed from the legislative payroll. As a convicted felon he could no longer hold public office. His replacement would be selected at a special election in January.

Surprised that his guilty plea had not resulted in probation instead of prison time, Lewis fought back. With a new attorney working the case, he began filing motions that would entangle the court in paperwork before, during and after his prison time. His new attorney was David Prosser, Jr. who would become Justice Prosser in an appointment to the Wisconsin State Supreme Court in 1998.

The extraneous motions frustrated Judge Doyle, who felt he had been quite lenient with Lewis' sentence. The final straw was Lewis' request for a court order to vacate his perjury sentence. Doyle rejected the request, and decided to release the court documents into the public domain, taking the extraordinary step of including some of the secret Grand Jury testimony.

Doyle made no announcement of the release. It would take two inquisitive reporters, Jack Anderson and Pat Rupinski, to discover them separately and by chance almost simultaneously.

With Lewis representing the West Bend area, it was natural that Anderson of the *West Bend News* had followed the case closely. Since learning of the Grand Jury probe in late 1979, he had poised himself to follow any tiny trail of information thrust into the light of day. To stumble upon the court documents while doing a routine follow-up on Lewis' many appeals, however, had been quite a surprise. He used the information he found in the file to begin a series of articles on the Lewis case beginning on November 4, 1980, leaving other newspapers to ponder why Doyle had released the file.

Meanwhile, Rupinski had started working on his own series of

Code Word Tikal...193

articles for the *Wausau Daily Herald* in direct response to a visit from Mike. The laser scientist had walked into the newspaper office one day and told the news reporter the whole, bizarre story. Mike's hope was to find in the press what he had not found in the courts: a sense of justice.

Rupinski was enthralled by the story, and, after getting approval from his editors, delved into the facts and details with enthusiasm. Prior to publication, the completed articles were reviewed by Rupinski's supervisors, both in Wausau and at the main offices for the newspaper's owner, Gannett. As if taking its cue from every management office involved in the case, from FBI Headquarters to the Justice Department, progress on the story stalled. Rupinski blamed the attorneys at Gannett who found the facts of the story too unbelievable for national publication.[77] To make it palatable for local publication, Gannett insisted on watering down the series.[78] It finally started running on November 6, 1980.

Rupinski was upset by the alterations and Gannett's legal fears that prevented the story from running nationally. He soon quit the *Daily Herald* and left Wisconsin to continue his reporting career in Alabama.

Mike was basically happy with the Rupinski and Anderson articles despite the omissions and changes. After the articles appeared, however, he began receiving phone calls. It was a mixture of praise and persecution. Although the majority was praise, a handful of mostly nameless callers chastised him. "People would say, you know, you did the wrong thing. They should have attacked Cuba. Cuba is communist," he remembered. They did not understand, did not see, that the plot embraced more than just anti-communist sentiment.

So, Mike took his case to local organizations for speaking engagements, warning of the anti-Semitic climate brewing in the nation, and the type of anti-government vigilantism that threatened the stability of the country. He received standing ovations and pats on the back, and nasty letters in the mail.

He wrote his own share of letters: letters to his fellow scientists warning of the similar dangers they faced as men and women of

---

[77] "He tells of plot to attack O'Hare Airport, Havana," by Hugh McCann, *Detroit News*, March 8, 1981.

[78] Interviews with Tom Burg and Mike Muckerheide.

knowledge, letters to Jewish leaders to warn them of the anti-Semitic tone of activism, and letters to Washington bureaucrats seeking answers and action. Again, there was a mixture of response.

"You should have taken the money," some people told him.

"Thank you for saving the country," said others.

And one: "You dumb shit! I wouldn't give you a nickel for your life."[79]

By chance, he met a reporter from Detroit and continued his crusade. Hugh McCann ran a major series in the *Detroit News* beginning in March of 1981. Mike also attempted to persuade some interest from TV news magazines, with little success. He was still caught in a time warp: neither laser weapons nor right-wing extremists meant much of anything to anyone yet. What was the big deal, anyway? No one had been hurt. The plot had remained a plot, never consummated.

The success of Mike's intervention had become his failure.

Finally, tired of the battle, Mike turned his back on his bitter past and turned to face his future.

The sentencing of Assemblyman Lewis signaled an important moment for him: he no longer had to hide behind a veil of secrecy. The truth of who he was and what he stood for were clearly defined: if not for others, at least for himself. He had survived his own doubts and was ready to take on the world. "I'm free again," he thought. "I'm a free man." He started setting his sights on what he wanted to do with his life. With intensity he began to pursue laser science with an unbridled passion.

Only one more hurdle would be thrown into his path. Bill Mark, his greatest supporter and financial backer for private research, died suddenly in October 1980. Some of the work the Mark Laser Company supported was non-medical in nature and the remaining Mark family did not wish to continue. Mark's widow offered Mike a supervisory position with A. Ward Ford Institute, but Mike declined. "I could not fly a desk," he said. "I could not."

The death of Mark "knocked the stilts out from under me," he admitted. With some of his closest contacts at NCTI also departing, he felt lost in Wausau and began venturing farther into the world beyond. "I just wandered around the system," he recalled. "Bumped here, bumped there, like a pool ball, and then it finally started to work for me."

---

[79] Letter from Mike Muckerheide files.

Code Word Tikal...195

During the early 1980s, he worked full-time for private medical laboratories in Milwaukee and Belleville, Illinois. With his daughter now entering high school, he resolved to commute from Wausau until she had completed her education there. For the weekly trek, he traded in his frumpy, old Pinto for a flashy Mazda RX7. Like his life, it signaled a change: he was leaving his broken past behind him.

There would be sporadic moments when some incident from the laser case would come back to haunt him. While on his weekly commute between Wausau and Milwaukee, before Iwen's car was confiscated and sold during the counterfeit case, Mike met the chemist on the highway near the tiny town of Ringle. It was the first time the two had seen each other in several months. Iwen pulled alongside Mike's car and paced him in his Mercedes. The two cars sped down the two-lane highway, side-by-side, fast and straight, each unwavering on the course. Finally, Iwen shot ahead, stopping at an overpass to wait. Seeing his opportunity, Mike hit the accelerator, his foot pressing down with all the frustration he felt over the fact that Iwen was still free. He sped past the Mercedes and into the distance, leaving Iwen in his dust.[80]

In some ways, it can be said that Mike kept on speeding ahead. While Iwen, the educated chemist, squandered his talents on questionable goals, entangling himself in the legal system, Mike, the self-made scientist, sailed on to loftier heights. After being courted by St. Mary's Hospital in Milwaukee, he decided in 1985 to sign on as their Laser Consultant and later Research Director. His primary responsibility was maintaining the hospital's growing inventory of lasers and helping to establish laser laboratories at affiliated hospitals throughout the country. He had many opportunities to leave Wisconsin, but he chose to stay, in part because his contract with St. Mary's allowed him to pursue his private research goals and he cherished the independence.

In 1986, Mike would file his fifth patent for a three-dimensional, laser driven display apparatus for image enhancement. It would be granted in 1989 and assigned to Setan Health Care Foundation, Milwaukee. Also assigned to the foundation was a patent for an improved laser pumping apparatus for increasing transmission efficiency, filed in 1989 and assigned in 1991.

Two more patents were assigned to Mike in the 1990s: a surgical

---

[80] Interview with Mike Muckerheide.

On the Laser's Edge...196

mask providing a barrier against infectious hazardous material such as HIV infected blood and contaminated body fluids (filed in 1988; assigned in 1991); and a laser driven optical communications apparatus for permitting verbally impaired or verbally disabled individuals, with limited physical mobility, to communicate with others, (filed in 1990 and assigned in 1993).

One of Mike's greatest personal goals was accomplished in January of 1986. With his love for rocketry never diminished, he watched his first NASA payload rocket into space on January 12 aboard Columbia Flight STS-61-C. Payload G449, called Project JULIE (Joint Utilization of Laser Integrated Experiments), was a part of NASA's Get Away Special (GAS) program, and it would make St. Mary's Hospital the first private hospital to sponsor a GAS payload. Some of his former colleagues back in Wausau would assist with the project. The 1.5 megawatt laser on that flight was part of a display at the National Air & Space Museum. Another part of the payload went on display at the Gallery of Flight at Mitchell Airport in Milwaukee.

Mike would keep his contact with NASA through the remainder of his career. He worked on several more payloads, and took particular delight in encouraging youth in some of the small payload experiments open to them through the GAS program. He connected with Arrowhead Union High School in Heartland for a student experiment involving the effects of laser stimulation on seed propagation in space. It flew aboard Discovery Flight STS-105.

He also worked with the youth at his alma mater, St. John's Prep, in Collegeville, Minnesota on a similar experiment.

Though certainly successful, Mike lacked the kind of funding he needed to accomplish all the things he wanted. As a result, a number of his successful ideas were never patented. Said his friend, Dr. Dudley Johnson, "I always think he felt frustrated in not being able to get enough done. Not enough time. Not enough support. Not enough financial backing..."

He learned to make do with what he had until it became a natural part of how he accomplished things. With his 3D imaging machine, he needed a certain piece of equipment he didn't have. Instead of acquiring the part, he developed his own by "cannibalizing it from a coffee pot heating element," Darrell Seeley noted, "which was just how Mike did things."

Dr. Johnson agreed. "He was a good organizer, pulling everything together to make things work."

Mike discovered early that finding money for military research was so much easier to get than for medical research. As a deeply religious man, however, he would continue to agonize over his work with laser weapons. In 1991, when the U.S. Government asked for his assistance in the Persian Gulf War, Desert Storm, he did so reluctantly. Once committed, he threw himself into the work. He was on call, traveling frequently, to help with top-secret research. He was proud to be helping his country; on the other hand, he wanted most to help heal people, not to help hurt them.

His military-related patents were sealed on security orders. His last patent application, likely to be sealed as well, dealt with using the energy of ordinary gunfire to power a laser, filed in 2003. If effective, it would help solve one of the major problems with laser weapons: creating compact, lightweight energy supplies.

He enjoyed another twenty productive years in a brilliant career after the laser case ended. Then, his wife's cancer returned, sending the family into another tailspin. Pat died on December 1, 2001. The following year, Mike lost his sister Mary to heart disease.

Mike's own cancer started sending out signals from his prostate while he was caring for Pat. The man who worked with doctors, but didn't trust most of them, refused to see any one of them until he was limping in pain. By then, the disease had advanced into his bones.

One of the laboratories Mike was then using was located at the Matrix Packaging Corporation in Saukville owned by his good friend Dee Willden. It was located on the upper level with only staircase assess, so Willden would have him hoisted up in his wheelchair on a forklift rack. Mike only visited a few times a month as his illness progressed, but worked on his last patent application there – capturing the power of gunshot – along with other projects.

One experiment that was making progress was the infusion of flavors into peanuts and coffee beans with lasers. "We would put the peanuts [and coffee beans] on a turntable. The laser would push the flavor into the center," Willden explained.

Seeley worked with him at the Matrix lab on Mike's final GAS payload that included an experiment called the Chemistry of Life. It essentially was delving into the creation of man; the first spark of life: a concept Mike was able to justify even as a religious man. The project was indefinitely sidelined with the 2003 Columbia explosion.

On the Laser's Edge...198

Willden was also deeply religious and the two men would occasionally have discussions about science and God. For Mike, the idea of light and how angelic beings might be transported on the beams of light was one that sparked his imagination. At times, it was coupled with his great sense of humor. The seemingly magical properties of holograms was an ideal way to accomplish their mischief making. "One time...we were going to put a hologram of the Savior on a cloud and see how many people would run. To see how many people would turn inside out," Willden said. "But we decided not to because we thought we'd cause a panic."

Another of Mike's whimsical experiments was the UFO Attractor. It was a large, metal, triple barrel contraption he located in his backyard off to one side of his view of Lake Michigan. Tom and I had immediately noticed the device on our first visit to interview Mike. With his usual dry humor he had told us: "I've only detected one set of footprints, and those were leading from the house."

Willden confirmed that it was a quasi-serious experiment. "It was a big play thing," he said. "It would have been a miracle if we had gotten something on it. It basically could have been done, but...we were just trying to stir up people."

They succeeded. They received wide news coverage for their efforts; though, not surprisingly, one of Mike's neighbors was not amused. "He didn't know what it was and he approached Mike, and I can't remember the answer Mike gave him, but he stayed away after that," Willden said.

Sometimes his research was done with and for his friends. These included systems for weighing particles in suspension as they fell, a lenless laser focusing device, and even a bug zapper. It was Dr. Johnson who wanted the bug zapper to keep grasshoppers away from his heirloom apple trees.

Towards the end, even as Mike's body failed him, his mind continued to swirl with ideas. Willden said he remained upbeat, talking about his work until the end.

Yet, my own phone interviews with Mike became very sporadic in the weeks leading up to early-July of 2003. I knew something was wrong by the tired sound in his voice. I pressed for a meeting so I could pick up a few stray articles he had found for me. When I saw him, I knew it would be for the last time. The fight was nearly gone, his anger replaced by painful resolve. A great life should not have to end this way; but then again, it was how it had started.

After that, he no longer called, and my messages to his answering machine went unanswered. Finally, in September I reached him through his daughter. I would talk to him once again, then the phone went dead.

It was obvious now that Mike was closing in around himself the people he knew and trusted the most: his family and his closest friends. He was saying his goodbyes.

Willden's family was among them. He and his wife, his son Marc and Marc's family made frequent visits. Marc's daughters would entertain him on the violin and piano. He liked classical music.

I received the call about his death three days before Christmas, December 22, 2003.

After his death, Mike's closest friends — Dee Willden, Darrell Seeley and Dudley Johnson — reflected upon the life of the self-made scientist. They all insisted that he knew much more than most people realized. "Mike was never prone to exaggeration," Seeley insisted. "People thought he was nuts, but it was all true. That was what was all the more intriguing about him."

"Mike was very unusual because he thought way outside the box. He was a visionary, seeing beyond what people could see," Willden explained.

"He was obviously very brilliant," Dr. Johnson added. "But his ideas were very practical, down to earth."

Mike's friends also realized, and accepted the fact, that there were parts of him for which he gave them little access. He was a man who kept his secrets. Some doors into his work and his past were forever closed.

"Much of what he worked on was classified, so he couldn't talk about it. He did a lot of work for the military, DOD, but I can't tell you anything about what that work was because he could never tell me," Seeley said.

Willden remembered: "I know, when he was out at the hospital [St. Mary's], he'd get a call and then I wouldn't see him for three or four days, and I learned never to ask him. He had a lot of things going on that nobody knew anything about. I would overhear things once in awhile. I never probed for anything. If he wanted to tell me, fine, because whenever I probed he'd get a little defensive."

As Marc Willden summed up, "There were just a couple of

crumbs that fell from the table. You didn't know what the whole meal was about, but you knew a little something."

Neither friends nor family really knew much about the laser case, either, because Mike seldom talked about it, or when he did, he didn't offer enough for people to truly understand. It was an odd case.

He was bitter, though; they knew that. And he told them often enough that he didn't trust the government because of the laser case: a lot of people had disappointed him. But he also let them know that there were people out there like Southworth, Page and Burg picking up the slack, and he would always be grateful for that.

There are indications that Mike worked for the government in laser research in more ways than people will ever know. He worked for them despite his disapointments because he continued to believe in his country and in the principles of the system. He likened the U.S. to a big, lumbering train that stayed on track regardless of itself and all its obstacles. It reminded him of something he had witnessed one day on a trip through his wife's hometown of Cologne, Minnesota. As he recalled: "A freight train was going by and I was parked at the crossing, waiting for it to pass. It was going like wild through the intersection. I'll bet it was going 80 miles-an-hour. I looked down towards the end of it and, my God, there was a big ball of dust rolling up past this tremendous line of cars. I realized they had a railcar on the ground. It was on the ground, riding between the rail! It came to the intersection where I was, and I realized this whole thing was going to derail. There was going to be a terrible train wreck.

"And there was a guy standing out there by the depot. He said, 'My God, there's a car on the ground! I gotta' get a message to those guys! They gotta' stop and look!'

"Then, all of a sudden, the car jumped up, hit the groove, landed back onto the rail, and kept on going. I couldn't believe it. Just like God picked that sucker up and set it back on its way."

Mike thought the United States was like that sometimes: one car off the track and dragging, the train never slowing and the car eventually turning itself upright and continuing on.

It was the perfect allegory for Mike's own life, too. Though the laser scheme left him questioning the American justice system, in his heart and soul he never stopped believing in his country, with all of its fickle flaws. What happened to him in the 1970s was a bleak diversion to a great career. So easily it could have led him down the path

Code Word Tikal...201

to his own derailment, leaving him bitter and broken; instead, he jumped the track, bounced back, and kept on going.

And what of the others in the laser case, Tom and I would wonder. What had happened to them in the intervening years?

James R. Lewis returned to West Bend after he was released from prison and became a lay minister for a nondenominational congregation. He went on to own and operate a health food store.

John Claussen had a successful, quiet career in investments. He and his wife became committed members of the Church of Jesus Christ of Latter Day Saints. He died in 2004 from cancer at the age of 57.

Luis Federico Fuentes was promoted to Brigadier General and served as defense minister of Guatemala, then executive director of the National Reconstruction Committee in the early 1980s. He died at age 53 in a single-engine airplane crash near Guatemala City on April 3, 1989 while taking an aerial photographer and an engineer on a chartered flight to photograph Guatemalan ranches.

Iwen, who claimed to know very little about Fuentes during the Grand Jury probe, went on to marry the Guatemalan's sister. After he served his prison time for counterfeiting, he began to settle into a more quiet life as the couple began a family. He would be described as a doting father whose severity of his early years mellowed. He continues to work part-time as a chemist at a vitamins lab in Central Wisconsin. He refused comment for this book, maintaining that the laser case had amounted to entrapment. He alluded only to errors of judgment regarding his militancy.

Thomas Stockheimer's luck ran out in 1995 when he was sentenced to 16 years in prison for assaulting two police officers the previous year. He got a head start at experiencing incarceration when he was arrested the same month and charged with mail fraud and conspiracy for selling bogus money orders through an organization he ran in Wisconsin called the Family Farm Preservation: a Posse Comitatus offshoot. One of Stockheimer's "customers" was Terry Nichols, the Oklahoma City bombing accomplice, who had obtained a worthless "Certified Fractional Reserve Check" in 1993 to pay off a bank loan.[81]

Frank Tuerkheimer continued his illustrious legal career. He was U.S. Attorney for the Western District of Wisconsin from 1977 until

---

[81] *The Terrorist Next Door: The Militia Movement and the Radical Right* by Daniel Levitas, Thomas Dunne Books, 2002.

On the Laser's Edge...202

1981, then entered private practice and teaching. He is currently a Habush-Bascom Professor of Law at the University of Wisconsin, where he was voted Teacher of the Year in 1999, and is of counsel for LaFollette, Godfrey & Kahn law firm. He was a visiting Professor of Law at Benjamin Cardoza School of Law in his native New York, and has been the author and co-author of several papers appearing in major U.S. law reviews.

Jerry Southworth retired from the Federal Bureau of Investigation and operates his own business advising companies in security matters.

Tom Burg also retired from the FBI in 1999, and from 2001 until mid-2004, was Security Advisor for the United States Figure Skating Association. A rail fan, he wrote the book *White Pine Route: The History of the Washington, Idaho and Montana Railway Company* published in 2003.

BIBLIOGRAPHY:
Abbreviated listing.

BOOKS

Anderberg, Major General Bengt and Wolbarsht, Dr. Myron L. *Laser Weapons: The Dawn of a New Military Age*. Plenum Press. 1992.

Barkun, Michael. *Religion and the Racist Right, Revised Edition*. University of North Carolina Press. 1997.

Corcoran, James. *A Bitter Harvest: Gordon Kahl and the Rise of the Posse Comitatus in the Heartland*. Penguin. 1991.

Eichenwald, Kurt. *The Informant: A True Story*. Broadway. 2001.

Hoffer, Eric. *The True Believer*. Perrenial Library. 1966.

Kessler, Ronald. *The FBI: Inside the World's Most Powerful Law Enforcement Agency*. Pocket. 1994.

Levitas, Eric. *The Terrorist Next Door*. Thomas Dunne Books. St. Martin's Press. 2002.

Maiman, Theodore H. *The Laser Odyssey*. Laser Press. 2000.

Roberts, Archibald. *The Republic: Decline & Future Promise*. Betsy Ross Press. 1975.

Walters, Jerome. *One Aryan Under God: Exposing the New Racial Extremists*. Pilgrim Press. 2000.

Wells, H.G. *War of the Worlds*. 1898.

Wisconsin Blue Books. 1972-1979.

ARTICLES

"Federal Bureau of Investigation History." http://www.fas.org

Freeman, D.J. *"History and Future of North Central Health Foundation."* May 16, 1973.

Robinson, B.A. "Christian Identity Movement". 1997-2000. Ontario Consultants on Religious Tolerance.

"Robert H. Goddard: American Rocket Pioneer." National Aeronautics and Space Administration. Goddard Space Flight Center. Greenbelt, Maryland.

"Watergate: Brief Timeline". Watergate.com.

Watergate: Chronology" . Watergate.com.

Various articles from *Wausau Daily Herald*, *West Bend News*, *Detroit News* & *Madison Capitol Times*.

INTERVIEWS:
Carol Barwick, David Barwick, Ron Beckman, Tom Burg, Tom Burgoyne, Diane Dalsky, Dr. Ellet Drake, Dr. D. J. Freeman, David Haskins, Mary Jane Uecker Hettinga, Dr. Dudley Johnson, James Lombardo, Jerry Madison, Mike Muckerheide, Jack Page, Barbara Peterson, Sandy Robarge, Susan and Mike Schreiner, Dan Sczygelski, Darrell Seeley, Juan Simon, Jerry Southworth, Steve Thompson, Frank Tuerkheimer, Janet Volpe, Dee Willden, Marc Willden.

OTHER:
Canadian patent office. Patents: 2121248; 1303762
Court documents. Case #79-CR-57. *United States versus James R. Lewis.*
    U.S. District Court. Western District of Wisconsin.
Court documents. Case #80-CR-35. *United States versus Albert Carl Iwen.*
    U.S. District Court. Eastern District of Wisconsin.
Marathon County Historical Museum Library files.
Marathon County Public Library.
U.S. patent office. Patents: U.S. 4,120,293; 4,161,944; 4,143,275;
    4,316,467; 4,799,103; 5,005,182; 5,012,805; 5,191,411
West Virginia Collection. Colson Hall Library. West Virginia University,
    Morgantown, WV.

ABOUT THE AUTHORS:

Sharon Thatcher is an associate editor and online editor for F&W Media/Krause Publications. She has a background in reporting and editing for small weekly publications, and dozens of her freelance articles have appeared in regional and special interest publications. She collaborated with the Merrill Historical Society for the publication *History of Merrill Wisconsin: The Jenny Years, 1847-1881*, which received a special commendation from the Wisconsin Historical Society. She was also editor of the book, *The Rangers' Reign: A Glimpse of Semi-Pro Baseball in the '50s* by Louis Paetsch and Michael Weckwerth.

Tom Burg was one of the FBI agents who played a pivotal role in the laser conspiracy case. He began his career in Mississippi, working with cases involving the White Knights of the Ku Klux Klan. He retired in 1999 as Senior Resident Agent in Wausau, WI. He is also a special interest writer and rail fan. Years of research culminated in the book, *White Pine Route: The History of the Washington, Idaho and Montana Railway Company* published by the Museum of North Idaho. The book was one of 17 nominated for the Idaho Library Association 2003 Book Award. He was Security Advisor for the United States Figure Skating Association from 2001 to 2004.

Sharon and Tom also collaborate on special interest photo CDs and photo books through their small publishing company, Merrill Publishing Associates.

Other books by Sharon Thatcher:
History of Merrill Wisconsin: The Jenny Years, 1847-1881
   Copyright 2000 Merrill Historical Society, Merrill, WI

Other books by Thomas E. Burg:
White Pine Route, The History of the Washington, Idaho & Montana Railway Company, Copyright 2003 Museum of North Idaho, Coeur D'Alene, ID

Other publications by Merrill Publishing Associates:
PHOTO CDs:
- *Burlington Route Steam Locomotives* Photo CD, 2005
- *Early 20th Century Rock Island Steam Locomotives* Photo CD, 2005
- *Soo Line Steam Locomotives* Photo CD, 2005
- *Walborn & Riker Catalog of Pony, Cob & Horse Pleasure Vehicles* CD, 2005
- *The Green Bay & Western Steam Era Locomotives* Photo CD, 2006
- *Milwaukee Road - Steam in the West Photo* CD, 2006
- *Davenport Locomotive Works Catalog* CD, 2007
- *Wisconsin Shortline and Logging Steam Locomotives* Photo CD, 2008

BOOKS:
- *Wisconsin Shortline and Logging Steam Locomotives, Photos from the Roy Campbell Collection,* 2009 (book reproduction of 2008 CD)
- *C&NW Steam Era in Wisconsin, Photos from the Roy Campbell Collection,* 2009
- *Chicago's Steam Suburbans, Photos from the Roy Campbell Collection,* 2007
- *Burlington Route Steam, Photos from the Roy Campbell Collection,* published in 2008 (book reproduction of 2005 CD)
- *Milwaukee Road Steam in the West, Photos from the Roy Campbell Collection,* published in 2008 (book reproduction of 2006 CD)
- *Soo Line Steam Locomotives, Photos from the Roy Campbell Collection,* published in 2008 (book reproduction of 2005 CD)
- *Early Twentieth Century Rock Island Steam, Photos from the Roy Campbell Collection,* published in 2008 (book reproduction of 2005 CD)
- *The Green Bay & Western Steam Era, Photos from the Roy Campbell Collection,* published in 2008 (book reproduction of 2006 CD)
- *Walborn & Riker Catalog of Pony, Cob & Horse Pleasure Vehicles,* published in 2008 (book reproduction of 2005 CD)

Available at: www.merrillpublishingassociates.com